Chinese Research Perspectives on the Environment, Volume 9

Chinese Research Perspectives on the Environment

Chinese Research Perspectives

General Editor

Steven L. Leibo (*Russell Sage College*)
Li Yang (*Prop Roots Program*)

VOLUME 9

The titles published in this series are listed at *brill.com/cren*

Chinese Research Perspectives on the Environment, Volume 9

Edited by

Liu Jianqiang

BRILL

LEIDEN | BOSTON

This book is the result of a co-publication agreement between Social Sciences Academic Press and Koninklijke Brill NV. These articles were selected and translated into English from the original《中国环境发展报告（2015）》(*Zhongguo huanjing fazhan baogao 2015*) with the financial support of the Alibaba Foundation.

Typeface for the Latin, Greek, and Cyrillic scripts: "Brill". See and download: brill.com/brill-typeface.

ISSN 2212-7496
ISBN 978-90-04-40156-3 (hardback)
ISBN 978-90-04-40157-0 (e-book)

Copyright 2019 by Koninklijke Brill NV, Leiden, The Netherlands.
Koninklijke Brill NV incorporates the imprints Brill, Brill Hes & De Graaf, Brill Nijhoff, Brill Rodopi, Brill Sense, Hotei Publishing, mentis Verlag, Verlag Ferdinand Schöningh and Wilhelm Fink Verlag.
All rights reserved. No part of this publication may be reproduced, translated, stored in a retrieval system, or transmitted in any form or by any means, electronic, mechanical, photocopying, recording or otherwise, without prior written permission from the publisher.
Authorization to photocopy items for internal or personal use is granted by Koninklijke Brill NV provided that the appropriate fees are paid directly to The Copyright Clearance Center, 222 Rosewood Drive, Suite 910, Danvers, MA 01923, USA. Fees are subject to change.

This book is printed on acid-free paper and produced in a sustainable manner.

Contents

List of Figures and Tables   ix

1 China's Growing Environmental Civil Society   1
    ZHANG Shiqiu

2 Lessons from "APEC Blue" about Air Pollution Control in China   24
    ZHAO Lijian

3 Controlling Aggregate Coal Consumption: Plans and Policies   36
    CHEN Dan, LIN Mingche, and YANG Fuqiang

4 Revisiting the Private Automobile Ownership Debate after 20 Years   54
    SHI Jian

5 Soil Remediation: Still a Long Road Ahead   60
    LIU Hongqiao

6 Progress in China's Environmental Legislation in 2014   72
    QIE Jianrong

7 The Financial Sector and Environmental Risks: Understanding the New *Environmental Protection Law*   83
    WANG Xiaojiang and WANG Tianju

8 Recent Developments in Environmental Criminal Justice   95
    YU Haisong and MA Jian

9 Environmental Information Disclosure Made Breakthroughs in 2013–2014   109
    WU Qi

10 China's New Urbanization Plan and Sustainable Consumption   120
    CHEN Hongjuan and CHEN Boping

11 The Imminent Threat of Tropical Viruses: Lessons from the 2014 Ebola Outbreak in Africa   132
    SHEN Xiaohui

12 Chinese React to Jack Ma's Hunting Trip to UK   142
    LIU Qin

13 Reflections on Outbound Investment by China's Mining Industry   150
    BAI Yunwen and BI Lianshan

14 A Case for Banning Illegal Timber Imports in China   163
    YI Yimin

15 Chinese Involvement in Brazil's Development: Massive Investments Bring Environmental and Cultural Challenges   176
    ZHOU Lei and Petras Shelton ZUMPANO

16 China's Role in Antarctic Marine Conservation   187
    CHEN Jiliang

17 Real-time Information Disclosure under Blue Sky Roadmap II   200
    Jointly released by the Institute of Public & Environmental Affairs; Society of Entrepreneurs & Ecology (SEE); the Institute of Environmental Policy and Planning, Renmin University of China; Friends of Nature; EnviroFriends; Nature University

18 An NGO Review of China's Carbon Market   206
    Greenovation Hub

19 Perverse Incentives: Electricity Generation through Waste Incineration and Renewable Energy Subsidy   225
    The Rock Environment & Energy Institute

Appendix I: Chronicle of Major Events in 2014   229
Appendix II: Green Book of Environment Evaluation of China's Environmental Performance in 2014   244
Appendix III: Air Quality Ranking for 2014 of Provincial Capitals and Central Government-Controlled Municipalities   262
Appendix IV: List of Environmental Protection Laws and Regulations Released in 2014   264
Appendix V: A Letter to the Government about Mandatory Protective Book Jackets   268
Appendix VI: Friends of Nature Suggestions for Amending the *Air Pollution Prevention and Control Law*   271

Appendix VII: Friends of Nature Continues to Urge Relaxation of Eligibility Requirements for Litigants in Environmental Public Interest Suits　275

Appendix VIII: A Petition Calling for Releasing for "Soliciting Public Opinions and Ensuring that the Newly Revised *Standards for Controlling Pollution from Residential Waste Incineration* by July 1, 2014"　277

Appendix IX: 2014 Winners of Ford Motor Conservation & Environmental Grants, China　280

Appendix X: 2014 Winners of Best Environmental Report Award in China　284

Appendix XI: 2014 Winners of UNEP Champions of the Earth Award　289

Appendix XII: 2014 Winners of the Green China · 2014 Environmental Protection Achievement Award　292

Index　294

Figures and Tables

Figures

2.1   Comparison of the trend of air pollutant emissions and that of economic and social development in the USA, 1970–2009   31
8.1   Composition of environmental pollution crimes from July 2013 to October 2014   96
8.2   Changes regarding the increase in criminal cases involving environmental pollution   98
8.3   Distribution of cases closed in the first instance from July 2013 to October 2014   100
8.4   The identity of criminals in the criminal cases involving environmental pollution from July 2013 to October 2014   103
10.1   China's sustainable consumption strategic framework   130
16.1   China's krill catch in the Antarctic   193
16.2   CCAMLR countries historical krill catch   194
16.3   2012 Ross Sea MPA Proposal   195
16.4   The trend in global marine fisheries   197
18.1   The framework of the Chinese carbon market evaluation system   218
II.1   Comparison of China's 10 largest river basins in 2013 by categories of water quality   246
II.2   Water quality of state-level monitored lakes and reservoirs during 2004–2013   248
II.3   Seawater quality of different coastal waters in 2013   250
II.4   Acid rain frequency nationwide during 2006–2013   254

Tables

2.1   Air pollution control measures adopted in Beijing during the APEC Meeting and their contributions to emission reduction   27
17.1   The results of the AQTI evaluation of 113 key cities for environmental protection   203
18.1   Basic information on the implementation plan for local pilot carbon trading (as of November 5, 2013)   210
18.2   Overview of the non-governmental evaluation system for the Chinese carbon market in 2013   219

19.1 Comparison of contribution rates of waste components in unit electric quantity 227
19.2 Comparison of carbon emissions under different waste disposal modes 228
II.1 Discharge of major pollutants in wastewater nationwide in 2013 244
II.2 Emission of major pollutants in exhaust gas nationwide in 2013 245
II.3 Generation and utilization of industrial solid waste nationwide in 2013 245
II.4 Comparison of China's 10 largest river basins in 2013 by categories of water quality 246
II.5 Water quality of major lakes and reservoirs in 2013 247
II.6 Water quality of state-level monitored lakes and reservoirs during 2004–2013 247
II.7 Total discharge of pollutants into the 4 largest seas in 2013 (in tons) 251
II.8 Numbers of cities in key areas meeting the standards of all pollutants 252
II.9 Monitoring points reaching the standards in functional zones of cities at or above prefecture-level in 2013 255
II.10 Monitoring points reaching the standards in functional zones of key cities of environmental protection in 2013 255
II.11 The amounts and areas of natural reserves in 2013 258
II.12 Production and growth rate of primary energy in 2013 260
IV.1 Laws and regulations 264
IV.2 Departmental rules 264
IV.3 Regulatory documents 265

CHAPTER 1

China's Growing Environmental Civil Society

ZHANG Shiqiu*

Abstract

The year 2014 is considered the 20th anniversary of the non-governmental environmental protection movement in China. As one of the hallmarks in the development of an environmental civil society, China's non-governmental organizations for environmental protection are continuously developing and growing, have participated in environmental governance at various levels, and have made irreplaceable contributions to China's environmental governance.

Keywords

good governance – social capital – non-governmental organization for environmental protection – civil society – environmental governance

More than 100 years ago, the British author Samuel Smiles maintained in his book *The Strength of Character* that the future of a country depends on the quality of its citizens' civilization—the education that people receive, people's foresight, sagacity and character—rather than its substantial national treasury, solid castles and magnificent public facilities.[1]

More than 100 years later, we can say that the environmental quality, ecosystem and the prospect of a sustainable development of a country hinge not only on the government's unilateral declaration of war against pollution, the political will in an iron-fisted control of pollution and the execution capability, but also the development, growth and maturation of China's environmental civil society, as well as the sophistication of the mechanism of interest check and balance and coordination that is based on this.

---

* Zhang Shiqiu, Professor at the College of Environmental Sciences and Engineering, Peking University, has engaged in research and teaching concerning environmental economics, environmental policy and management for a long time.
1 Samuel Smiles, *The Strength of Character* (Beijing Library Press, 1998).

The year 2014 was the 20th anniversary of the establishment of non-governmental organizations for environmental protection in China. Since the reform and opening up, especially in the past 20 years, China has witnessed great social and economic changes and a substantial improvement in the people's living standard, which have also been accompanied by a number of problems, including environmental pollution, ecological destruction, global warming, resource depletion and a sharp decline in biodiversity. The general public has shown increasing concerns about environmental problems, especially those concerning human health and caused by environmental problems; the voice for good governance of the environment has been running high, so non-governmental organizations aimed at environmental protection have emerged and the non-governmental movements have rapidly developed in China.

During the past 20 years, as one of the hallmarks in the development of an environmental civil society, non-governmental organizations (NGOs)[2] for environmental protection in China—also called association organizations for environmental protection, social organizations for environmental protection—started from scratch, have rapidly developed and grown. They are the NGOs which were established first and are enjoying the fastest development and the greatest social influence in China. NGOs for environmental protection display the developmental trends and characteristics of the development of an NGO in China and are also important examples of the changes in China's social organizations and their operational modes.

With progress in the Chinese society, people have shifted their focus from whether NGOs, especially NGOs for environmental protection, should be listed as "forbidden areas" and "sensitive" topics to how to effectively leverage the roles of NGOs for environmental protection, so as to promote the protection of the ecological environment and resources, good governance of China's ecological and environmental protection, and boost the development of China's

---

[2] An NGO has the following main characteristics: non-governmental, non-profit, organizational, autonomous, voluntary and public welfare-oriented. The World Bank defines it as follows: An NGO is a professional organization under independent operation which consists of volunteers with the same intention, has a stable organizational form and leadership structure, and is not designed to make profits and is detached somewhat from governmental bodies and business organizations (World Bank, "Involving Nongovernmental Organizations in Bank-Supported Activities," Operational Directive 14, Washington DC: World Ban k, 1989, p70). An NGO for environmental protection is a social organization which is organized for the special purpose of protecting the ecological environment in addition to sharing common characteristics with an NGO. An NGO for environmental protection is an independent social organization which is the third sector between the government and the general public, and enterprises.

environmental civil society. We can even expect that the environmental field will very likely become one of the most important fields for generating a mature civil society and thus accumulate the necessary social capital for China's long-term sustainable development. Besides the characteristics of a civil society, a mature civil society in the environmental field presents at least the following characteristics: (1) it has the character and complexity of natural and humanistic care; (2) it shows care for a long-term sustainable development of the society; (3) it unremittingly enhances the environmental awareness and places equal emphasis on environmental ethics and high self-discipline and conscious environmentally-friendly acts; (4) it rationally seeks and safeguards the environmental rights and interests of citizens and society; (5) it pays equal attention to seeking and guaranteeing rights and interests and assuming environmental responsibility; (6) there is a sense of participation in and execution of social public issues, and the resulting social network of a citizen's autonomy and favorable interaction among multiple social players; (7) there are the social organizations, operational mode and system through which social conflicts are resolved by multi-party cooperation, check and balance and game playing.

The omens and reasons for the above expectations may be found in the revision and implementation of the *Environmental Protection Law* in 2014. Among the great events involving environmental protection in 2014, the revision and adoption of the *Environmental Protection Law* is certainly the most impressive. The very strict *Environmental Protection Law (Amendment)* was adopted during the Standing Committee of the National People's Congress on April 24, 2014 which became effective on January 1, 2015. It, for the first time, explicitly grants subject qualification for public interest litigation to the social organizations which are registered with the civil affairs department of the people's government above the level of the city divided into districts, have specialized in public benefit activities relating to environmental protection for more than five consecutive years and are not recorded as having violated the law.[3] Meanwhile, the new *Environmental Protection Law* also sets forth the legal provisions for the disclosure of information and public participation. According to Article 53 of the *Environmental Protection Law*, citizens, legal persons and other organizations enjoy, according to laws, the right to obtain environmental information and to participate in and supervise environmental protection. The competent departments for environmental protection within the people's governments at various levels and other departments responsible

---

3 The *Environmental Protection Law of the People's Republic of China* officially took effect on January 1, 2015.

for environmental protection, supervision and management should make public the environmental information and improve the public participation procedures according to laws; they should also make it convenient for citizens, legal persons and other organizations to participate in and supervise environmental protection.[4] On January 7, 2015, the *Interpretation of the Supreme People's Court of Several Issues Concerning Applicable Laws for Hearing the Cases Involving Environmental Civil Public Interest Litigation* officially came into force; the Supreme People's Court, the Ministry of Civil Affairs and the Ministry of Environmental Protection jointly issued a circular to specify the requirements for carrying out the environmental civil public interest litigation system and provided a more comprehensive legal guarantee for the participation of social organizations in environmental protection.

Moreover, in order to ensure that the NGOs for environmental protection have a deeper understanding of the current environmental situation and environmental policy and enhance the capability of the NGOs for environmental protection to effectively participate in environmental affairs under the new situation, the Ministry of Environmental Protection held the 2014 training workshop for social organizations for environmental protection in Nanjing, in November 2014. The persons in charge of 55 social organizations for environmental protection from 24 provinces, autonomous regions and municipalities directly under the Central Government nationwide attended that training workshop. That was the first time that the competent national department for the environment took the initiative in calling NGOs for environmental protection for training. This shows that the competent department for the environment has realized the importance of communicating and cooperating with the NGOs for environmental protection. Moreover, it also indicates that the competent department places high hopes on the roles of the NGOs regarding environmental protection.

This year's green book on the environment continues the previous years' discussions about the major environmental issues and environmental policies. Each article and each case reveal the important contributions of a developing environmental civil society in promoting China's environmental management and sustainable development. Of course, in China, which is undergoing rapid changes, transition and transformation, like other players, the NGOs for environmental protection are also facing a number of challenges and problems as well as many opportunities for development. These challenges and

---

4 The *Environmental Protection Law of the People's Republic of China* officially took effect on January 1, 2015.

opportunities come not only from the outside world, but also from the NGOs for environmental protection.

This chapter reviews the development of China's NGOs for environmental protection and further discusses the roles and status of the NGOs for environmental protection, their challenges and the opportunities open to them.

1   Review of the Development of China's Non-governmental Organizations for Environmental Protection and Environmental Civil Society

On June 5, 1993, several open-minded intellectuals, including Liang Congjie, member of the National Committee of the Chinese People's Political Consultative Conference and mentor at the International Academy for Chinese Culture, and Yang Dongping, professor at the Beijing Institute of Technology, held the first non-governmental symposium on the environment at Beijing's Linglongyuan Park—Linglongyuan Meeting. That meeting marked the official establishment of the Friends of Nature.[5] After setbacks, on March 31, 1994, that organization was registered with the Ministry of Civil Affairs and officially became a secondary body under the International Academy for Chinese Culture. Its full name is Green Culture Branch of the International Academy for Chinese Culture—Environmental Research Institute of the Friends of Nature, Chaoyang District, Beijing, which is abbreviated to the Friends of Nature. Mr. Liang Congjie served as its president. The Friends of Nature became the first NGO for environmental protection registered with the Ministry of Civil Affairs in China. The year 1994 is widely believed to be the first year for NGOs in China's environmental field.[6] Subsequently, the Global

---

5 Source: website of the Friends of Nature. http://www.fon.org.cn/index.php/Index/post/id/26.

6 In fact, the NGO for environmental protection registered and established at the earliest in China is Saunders's Gull Conservation Society of Panjin City in Liaoning Province. It was registered on April 18, 1991. It was initiated by Liu Detian who was a journalist of the *Panjin Daily* (source: China's Life History of NGOs for Environmental Protection. *Southern Weekend*, 2009-10-08). Furthermore, some people consider 1978 as the first year for NGOs in China's environmental field because the Chinese Society for Environmental Sciences was established in 1978; however, it has basically an official background, so it is not regarded as a modern NGO in the industry. Properly speaking, the Chinese Society for Environmental Sciences, the China Wildlife Conservation Association and the Chinese Society for Sustainable Development can be called GOVNGOs because they are NGOs sponsored by the government or in which the personnel appointed by the government hold the main posts, they have clear administrative ranks and their volunteers are mainly mobilized by the government to take part in them.

Village of Beijing and the Green Earth Volunteers were established in 1996. The Friends of Nature, the Global Village of Beijing and the Green Earth Volunteers were the benchmarks and leaders of China's NGOs for environmental protection in the 1990s. Afterwards, the NGOs for environmental protection were set up by non-governmental forces nationwide. That was the beginning of the NGOs for environmental protection, environmental public participation and development of an environmental civil society.

According to the *Blue Book on the Developmental Status of China's Non-governmental Organizations for Environmental Protection* released by the All-China Environment Federation, as of October 2008, there were 3,539 NGOs for environmental protection nationwide, among which 1,309 environmental protection organizations were established by the government, accounting for 37%, and 508 were grass-roots ones, accounting for 14%. According to the *2012 Statistical Report on Social Service Development* released by the Ministry of Civil Affairs, as of the end of 2012, there were 7,881 ecological environmental organizations nationwide, including 6,816 social bodies. During the five years from 2007 to 2012, the NGOs for environmental protection grew annually by an average of more than 30%.

During 20 years of development, the NGOs for environmental protection have been on the rise, with increasing members and a rapidly growing influence. The NGOs for environmental protection have exerted a profound impact on a sequence of major environmental events, such as the protection of the Tibetan antelope, the resettlement of Shougang Group, the construction, postponement, even suspension of major water conservancy projects like Nujiang River hydropower and Songhuajiang water conservancy, and Xiamen's PX project, Panyu's waste incineration plant project, molybdenum copper factory project in Shifang County of Sichuan Province and pollution discharge from Wangzi paper mill in Qidong of Jiangsu Province.[7] This further shows that environmental issues have become economic, social, livelihood, even political issues.

1.1 Development Stage and Roles of the NGOs for Environmental Protection in China

The *Blue Book on Development Status of China's Non-governmental Organizations for Environmental Protection* released by All-China Environment Federation in 2006 summarizes the characteristics of the development of

---

7 Zhang Xiaoran, "Declare War on Pollution: the Growing NGOs for Environmental Protection," *China Industrial Economy News*, 2014-06-05.

China's NGOs for environmental protection, and divides the course of their development into three stages: the stage of generation and rise (1978–1994), the development stage (1995–2002) and the expansion stage (2003 and beyond).

According to the Blue Book, China's NGOs for environmental protection started from May 1978 and were marked by the Chinese Society for Environmental Sciences which was initiated and established by the government department, while the establishment of the Friends of Nature in 1994 marked the establishment of NGOs for environmental protection.

During the years 1995–2002, China's NGOs for environmental protection proceeded from species conservation, to which the general public paid attention, and carried out various publicity activities including the Yunnan snub-nosed monkey and the Tibetan antelope conservation actions to increase social attention to the issues concerning endangered species and the environment. Moreover, the Global Village of Beijing cooperated with Beijing Municipal Government in starting pilot work on green communities as from 1999. Afterwards, China's NGOs for environmental protection gradually expanded from animal conservation to environmental awareness publicity and practice in environmentally-friendly behaviors, and expanded continuously into cities and communities, enhancing public awareness about environmental protection, including garbage classification, bird conservation, tree planting, conservation of endangered species, creation of green communities and environmental awareness publicity.

The year 2003 represents a period of transformation for the development of China's NGOs for environmental protection, during which the efforts made by single organizations in public environmental awareness and behavioral change in carrying out environmental publicity and promoting the conservation of particular species at the initial stage, were gradually developed into organizing the general public for participating in environmental protection and bringing attention to and involvement in major social development and decision-making issues through alliance and cooperation. The iconic actions included debating on the development of Nujiang River hydropower and the 26°C air-conditioning action initiated by NGOs for environmental protection, such as the Friends of Nature. More importantly, at the current stage, the NGOs for environmental protection have initiated public participation and have shifted their focus to offering advice and suggestions about the national environmental cause, exercising social supervision, safeguarding the public environmental rights and interests, and stimulating sustainable economic and social development. The above changes have also marked the expansion of the space for the activities of China's NGOs for environmental protection to actions aimed at safeguarding rights. Many NGOs for environmental protection have incorporated the safeguarding of public environmental rights and interests into their action programs or topics for activities.

1.2 *The Roles of the NGOs for Environmental Protection*

The activities carried out by the NGOs for environmental protection in the past 20 years have enhanced public awareness about environmental protection and have influenced the environmental behavior of the public. More importantly, the NGOs for environmental protection have become the indispensable players in carrying forward China's cause of environmental protection; they make up for the deficiency of the government and enterprises in providing environmental public services and play the roles of offering advice and exercising supervision. The common understanding of the roles which China's NGOs for environmental protection play can be summarized in the following eight aspects.

1.2.1 Effectively Promote Education about and Publicity Concerning Environmental Knowledge

Almost all of the NGOs for environmental protection have worked on taking the responsibility for popularizing knowledge of the environment since their inception. Their efforts lie in launching various environmental protection initiatives and practical activities, holding lectures and carrying out training and publicity to enhance the environmental awareness of the general public and various players, spreading knowledge about the environment, convening seminars and symposiums in various forms in order to boost the communication among the decision-makers, enterprises, social organizations and the general public, and further to promote public participation in environmental public affairs.

1.2.2 Push Forward and Promote Public Participation in the Field of Environmental Protection

Public participation is the essential part of a good governance of environmental protection. The *Environmental Impact Assessment Law* implemented as of 2002 specifies the legal status and role of public participation. Public participation is reflected in public involvement in environmental protection activities and practice in environmentally-friendly behavior. More importantly, more and more signs show that public participation has expanded from passive participation to active offering of advice and supervision over the government's policy-making and environmental behavior, and the enterprises' environmental performance and unlawful acts.

1.2.3 Implement Specific Projects for Protecting the Natural Resources and the Environment

Efforts have been made to accelerate wild animal and biodiversity protection, rational application of agricultural chemicals, especially pesticides and fertilizers, purification of the water and protection of the quality of the water in drainage basins, air pollution control, prevention and control of desertification, return of farmland to forests or grassland, prevention of water and soil erosion and ecological restoration.

1.2.4 Establish Non-profit Foundations for Environmental Protection, and Provide Support and Help for Non-governmental Environmental Protection Activities

For example, the SEE Foundation, highly active in recent years, is an NGO for the prevention and control of desertification sponsored by nearly 100 Chinese entrepreneurs. It is dedicated to the control of desertification and ecological protection in the Alxa area while stimulating Chinese entrepreneurs to assume more social responsibilities and offer fund support in various ways for the environmental protection activities promoted by the NGOs for environmental protection.

1.2.5 Boost the Production and Promotion of Environmentally-friendly Products, Green Supply Chain Management and Consumers' Assumption of Responsibility for the Environment

More and more NGOs for environmental protection are taking advantage of their capability for mobilizing social resources, utilizing the research conducted by them, experts, scholars and practitioners in relevant fields and publicity platforms in order to promote the R&D, production, circulation and consumption of environmentally-friendly products, and pushing forward environmentally-friendly behavior in the entire society through green production, green supply chain management and green consumption.

1.2.6 Carry Out Policy Research and Offer Advice and Suggestions Regarding Decision-making

A growing number of NGOs for environmental protection have turned their attention to research on macro-environmental policies while continuing to popularize and disseminate knowledge about the environment. For example, the Friends of Nature has released *China's Environmental Development Report* each year since 2009. This report is prepared from the perspective of China's non-governmental environmental protection organizations; it bluntly addresses hot issues of society, and provides an in-depth analysis. There are many other research projects on policies and on the environmental situation which attract extensive attention from society, and they are coordinated and organized by the NGOs for environmental protection; for example, with the coordination and organization by the Natural Resources Defense Council (NRDC), and with cooperation from the think tanks of the Chinese Government, scientific research institutions and universities and 19 influential organizations from industrial associations, the project of "A Study of Total Quantity Control Plan and Policy for Coal Consumption in China" was launched in October 2013, which offers many policy suggestions regarding policies and operational measures for setting the control objectives for the national total quantity of coal consumption, and implementing the roadmap and action plan; it is also dedicated to helping China achieve multiple objectives of resource conservation, environmental protection, climate change and sustainable economic development.

1.2.7 Carry Forward Information Disclosure and Public Interest Litigations, and Offer Legal Assistance for Environmental Pollution Victims

For example, the China Water Pollution Map, released by the Institute of Public & Environmental Affairs, and the Blue Map, which comes from the Institute and has boasted a large number of users, have become the important channels for the ordinary people to obtain information regarding air and water pollution. Over the years, the Center for Legal Assistance to Pollution Victims, established by China University of Political Science and Law in October 1998, has provided legal assistance to pollution victims; more importantly, it cooperates with the NGOs for environmental protection, including the Friends of Nature, and plays an important role in promoting environmental public interest litigations.

1.2.8 Enhance International Communication in Environmental Protection

During the initial development of China's non-governmental environmental protection organizations, there were homegrown grass-roots organizations and international NGOs which also helped China develop. The NGOs for environmental protection carry out international communication activities in various ways in order to understand and learn from the experience of international NGOs and obtain information, funds and technical support; they get involved in important international environmental issues, such as climate change, along with improvement of their own capability and attention to international topics. In recent years, China's NGOs for environmental protection have often been present and their voice has frequently been heard in important international meetings on environmental issues.

1.3 Several Crucial Changes in the Development of China's NGOs for Environmental Protection

China's NGOs for environmental protection have been late to develop, but they underwent many major changes in the short period of 20 years, as rapid economic, social and institutional changes occurred in China, and environmental and ecological problems, in particular, became increasingly severe. The crucial changes can be summarized as follows.

1.3.1 From a Small Number to Great Diversity

First, China's NGOs for environmental protection were only the Friends of Nature, the Global Village of Beijing and Green Earth Volunteers until the middle and later periods of the 1990s; today, about 1,000 NGOs for environmental protection, which are fully formed and registered by non-governmental forces, are available.

Second, geographically, the NGOs for environmental protection were previously concentrated in Beijing, Shanghai, Tianjin and the eastern coastal areas; at present, there are NGOs for environmental protection with on-going activities in almost every province and region.

Third, from the perspective of focus and scope of their activities, the NGOs for environmental protection have turned their attention from early environmental knowledge publicity, garbage classification and community practice to the current public participation, information disclosure, the safeguarding of environmental rights, policy suggestions, supervision over illegal environmental behavior, the enterprises' environmentally-friendly behavior and promotion of assuming more responsibility regarding the environment, green supply chain management, green consumption and green travel. This means that the participation of China's NGOs for environmental protection and that of the general public have greatly changed, from focusing on and caring for the environment to practicing environmentally-friendly behavior, even social supervision. It is particularly important that the attention from the NGOs for environmental protection and from the general public is paid not only to the protection and safeguarding of private rights relating to them, but also to the common environmental issues—the environmental issues concerning indirect interests. Citizens in a society should show interest in altruistic common issues along with their own rights. This is an important indicator for measuring the social moral standard and the degree of the social progress.

Fourth, with respect to the make-up of their personnel, in the early period, the NGOs for environmental protection mainly consisted of intellectuals and social activists keen on environmental protection, while currently they are made up of environmental protection enthusiasts, learned experts and scholars, media practitioners, students, ordinary citizens and entrepreneurs.

Fifth, organizationally, the NGOs for environmental protection included international NGOs, NGOs registered with the Ministry of Civil Affairs and with members available nationwide, local NGOs, specialized NGOs dedicated to certain issues and policy issues, comprehensive NGOs and many NGOs for environmental protection that developed out of student organizations.

Sixth, in the past, NGOs for environmental protection merely depended on their own funds or obtained support from a small number of international foundations; currently, sources of funds have become diverse, including support from domestic and foreign foundations, public welfare fund sponsorship from enterprises, individual donations and fund support from the government for purchasing public services from the NGOs for environmental protection.

Seventh, the NGOs for environmental protection have become diverse in terms of participation mode and information channels. The traditional modes for environmental participation by citizens include letters and visits for lodging complaints concerning environmental issues to higher authorities, the offering of suggestions to the People's Congress and the Political Consultative Conference, the filing of administrative arbitration to environmental administrative departments at various levels, subsequent public participation in environmental impact assessment of important investment projects, hearing of major environmental public policies, plans and strategies for development, administrative litigation against inactions or injustice of the environmental administrative departments, civil or criminal lawsuits for seeking legal means to solve pollution problems, and involvement of the NGOs for environmental protection and experts in government decision-making. With the development of the Internet and information communication technology, public environmental participation has become more active, and collective actions for environmental protection which exert a high social influence can rapidly take shape by virtue of the modern information communication technology, for example, in the case of Xiamen's PX incident, the environmental demonstration in Pengzhou of Chengdu City, the environmental mass incidents in Shifang and Qidong. Furthermore, individuals and NGOs for environmental protection take full advantage of WeMedia to release various types of information regarding the environment, acknowledge environmental concerns and reveal illegal environmental behavior; these phenomena have rapidly grown, posing a huge supervision pressure on lawbreakers and regulators.

1.3.2 From Making Good on Omissions and Deficiencies to Playing the Role of Leader

In recent years, the role of China's NGOs for environmental protection has gradually changed—at the early stages, the NGOs acted as a pusher in executing the government's environmental policies, made good on omissions and deficiencies, played a supporting role in the fields inaccessible to the government, carried out environmental protection publicity and served as the environmental guardian and representative of public environmental interests; later, they became an important social force exercising environmental supervision and a pressure group which could participate in the government's environmental decision-making, influence the environmental behavior of enterprises and the general public, and supervise the government's law enforcement and enterprises' environmental behavior.

During the present stage, environmental protection publicity and education remains the focus of the NGOs for environmental protection and their field of activity; however, they have reached out to the following several fields.

First, they supervise the environmental behavior of the government and enterprises. This supervisory role is continuously played and reinforced by means of social media and WeMedia; for example, the dissemination of PM2.5 prevention and control and the influence on it which started in 2011, the supervision over inadequate enforcement of environmental protection laws at the local level, and the supervision over and reporting of the surreptitious discharge of pollution by enterprises have exerted effective pressure of public supervision.

Second, they participate in decision-making in the environmental and developmental fields, which embodies public participation in and influence on major environmental public affairs. This includes the following aspects: participate in making national environmental protection laws, regulations and policies and exercise law enforcement supervision over the administrative organs by putting forward suggestions, addressing inquiries and holding hearings; call for and practice disclosure of environmental information; participate in developing plans, strategies and policies at various levels and environmental impact assessment of the projects. This not only influences and boosts major environmental legislation and decision-making, but it also effectively promotes sustained attention from the general public, enterprises and the government to environmental and developmental issues. For the amendments to important environmental laws and regulations, including China's *Environmental Protection Law* and *Air Pollution Prevention and Control Law* in 2014, many NGOs for environmental protection expressed very systematic and professional revisionary opinions, and mobilized the whole society to participate in that process. Some opinions were reflected in the amendments of the final draft; the most typical example was the discussion about and debate on the subject of public interest litigations involving environmental protection in the amendment to environmental protection law.[8]

Third, they push forward the disclosure of information regarding the environment, and provide the general public with that kind of information. China's NGOs for environmental protection independently make their own efforts or participate in the consulting publicity activities organized by relevant government departments, improve their own capability of social mobilization and organization and their own professional capability to bring about massive information disclosure, and prepare and promote policy research reports; for example, the annual green book of the Friends of Nature, the China Water Pollution Map Database released by the Institute of Public & Environmental Affairs as from 2006, the visual China Water Pollution Map, the release of air quality information as from 2011, the release of enterprise pollution discharge information, and the Blue Map, developed and applied in 2014, provide the general public with relevant high-quality information about the environment by adopting modern information technology; more importantly, they promote the disclosure of enterprise pollution discharge information and lay the necessary foundation for public participation in environmental protection.

Fourth, they propel environmental public interest litigation, and protect environmental rights and interests of the public. China was late to practice environmental public interest litigation, but this litigation has started exerting influence, and this kind of influence will increase along with the implementation of the new *Environmental Protection Law*. The recent years have witnessed some environmental public interest litigation cases with a great social influence; for example, six enterprises in Taizhou City, Jiangsu Province illegally

---

8 Regarding the *Amendment to the Environmental Protection Law (Draft)* (draft amendment), the first draft did not incorporate public interest litigation, thus triggering a heated debate. The second draft for deliberation stated that the only organization qualified for filing public interest litigation was the All-China Environment Federation, so many public interest organizations for environmental protection lodged complaints to higher authorities. In the third draft for deliberation in October 2013, the subject of public interest litigation was changed into "the national social organizations which are registered with the civil affairs department under the State Council according to laws, have specialized in public benefit activities relating to environmental protection for more than five consecutive years and enjoy a sound reputation". The final draft released in 2014 grants the subject qualification for public interest litigation to "the social organizations which are registered with the civil affairs department of the people's government above the level of the city divided into districts, have specialized in public benefit activities relating to environmental protection for more than five consecutive years and are not recorded as having violated any laws."

dumped waste acid, polluting a tributary of the Yangtze River, so a court ruled that those enterprises should make compensation of up to 1.6 billion yuan by the end of 2014, and some persons in charge were ordered to assume the corresponding criminal liability. In January 2015, after the new *Environmental Protection Law* was implemented, the Friends of Nature immediately initiated the Environmental Public Interest Litigation Support Fund. A number of public interest litigations supported by this fund, including the case of the ecological destruction in Nanping of Fujian Province and the case involving the ecological destruction of the mangrove forest ecology in Hainan, were placed on file for investigation and prosecution.

1.3.3 From Concerning Local Affairs to Participation in International Issues

The early development of China's NGOs for environmental protection was basically fostered at the local level. This is because China had severe environmental problems and there were restrictions on the focus on, the professional competence of and the international vision of the NGOs for environmental protection. Gratifyingly, China's NGOs for environmental protection appeared on the international arena as from 2000; for example, more than 150 representatives of 30 real NGOs for environmental protection from China attended the Johannesburg Conference on Sustainable Development in 2002.[9] In response to the issue of climate warming which has attracted extensive attention in recent years, China's NGOs for environmental protection have voiced their opinions in international conferences while practicing greenhouse gas reduction in China; for example, the position paper issued by the China Nongovernmental Climate Change Action Network at the UN Climate Summit 2014 aroused great widespread attention.

2 Environmental Civil Society/NGO, the Essential Part of Good Governance of the Environment

The government, enterprises and the general public (the civil society) are three players which are different but interrelated, they support, even restrict each other in a modern society. The environmental quality and the natural capital base have a bearing on the basic public services in a society, so the government is certainly duty-bound to ensure that its citizens enjoy a safe environment;

---

9 Huang Haoming, "The Roles of Non-governmental Organizations in Sustainable Development," *Xuehui*, 2004 (2).

however, in its practical operation, the government fails to effectively provide public services—government failure occurs, and the market is unable to offer effective public services—there is market failure—due to a lack of proper intervention and the enterprises' pursuit of profits. As social organizations composed of citizens, the NGOs constitute one of the important players in social resource allocation; they can effectively make up for government failure and market failure, and supervise the government and enterprises to assume their basic environmental responsibilities. Therefore, the NGOs/social organizations are also often called "the third sector in a society" in parallel with the government and enterprises. The entire range of social organizations, including the NGOs, is also regarded as "a civil society." A modern society is precisely a society where there are three forces—the market, the government and the civil society.[10]

2.1 Changes in the Environmental Management/Governance System and the Development of a Civil Society are Indispensable

China has witnessed rapid economic and social development; however, environmental pollution and the resulting problems concerning food safety and human health, resource depletion, energy security and problems regarding climate change have become the important ones which have drawn extensive attention at home and abroad. Moreover, environmental problems have exceeded the environment itself and have become social and political problems and the important factors affecting China's social stability and the harmonious development of the country. China's environmental management system and management institutions, especially the changes in the national environmental governance mode, are crucial for permitting China to switch to a green economy and to go down the road towards sustainable development. The bone of contention no longer involves whether change should be made; instead, it focuses on whether managers can conform to the situation and how they should adapt to the change of situation, whether they carry out top-down change in line with the situation or passively launch a change under pressure.[11]

At present, the environmental problems have become social and political problems, while mass incidents caused by environmental pollution are obviously increasing, and related public events frequently occur; the actions

---

10   Huang Bingyuan, "Transformation, Changes and Future Roles of Non-governmental Organizations (NGOs) for Environmental Protection," *Sina Blog*, blogger Jiutian Fangzhu. 2011-08-07.

11   Zhang Shiqiu, "Green Good Governance and Social Development," *Green Leaf*, 2013 (5).

and administrative effectiveness of the government in providing a good environment, controlling pollution, guarding against environmental health risks and safeguarding the basic environmental rights and interests of citizens are increasingly questioned. Mass disputes, interest disputes and qualms from the general public highlight the general public's doubts about the government due to its inaction and level of credibility, and also show the dissatisfaction with the government's failure to assume its environmental responsibility and its management/governance capacity; they also reflect the society's rational expectations and requirements for the government's improvement in environmental public management/governance and environmental public services.[12]

In the face of increasingly complicated public social affairs, there is an urgent need for China to change from the previous monistic structure to a pluralistic governance, from the traditional state-led governance to collaborative governance exercised by the state, the society and the citizens. The modernization of the national governance breaks through the restrictions from the inherent "government-centered theory" and aims at the reconstruction of the original national governance subject and governance mechanism. As a bridge between the state and the society, civil social organizations play an indispensable role in promoting the modernization of the national governance system and governance capacity and in achieving the overall objective of comprehensively intensifying the reform.[13]

The third adjustment mechanism based on the theory of the civil society—social adjustment mechanism, namely, embodiment by non-governmental and non-market means, such as public opinion, social morality and public participation[14]—or a civil society necessarily becomes the important part of a modern national environmental governance system. Yu Keping believes that a civil society can be considered the aggregate of all non-governmental organizations or non-governmental relationships outside the national or governmental system and the market or enterprise system; it is the non-governmental public field beyond the official political field and the market economic field.[15]

---

12   Zhang Shiqiu, "Green Good Governance and Social Development," *Green Leaf*, 2013 (5).
13   Li Huadong, "A Brief Analysis of the Roles of Civil Social Organizations in the Modernization of National Governance," *New West*, 2015 (2).
14   Liao Jiankai, "An Exploration of the Legality of Domestic Environmental Non-governmental Organizations," *Environmental Science and Management*, 2005 (3).
15   Yu Keping, "China's Civil Society: Concept, Classification and Institutional Environment," *Social Sciences in China*, 2006 (1).

2.2 *An Environmental Civil Society Promotes the Good Governance of the Environment and Enhances the Modernization of Environmental Governance*

Conflicts of interest caused by environmental problems are escalating, so our original extensive management means have to become fine, flexible and must stress interest coordination; only when an iron hand is combined with good governance can the economic growth pattern be fundamentally changed, the social, economic and environmental coordinated development be enabled and a win-win outcome in industrial production, life and ecology be achieved.[16]

As early as 1992, the United Nations clearly stated in the Agenda 21, which it adopted, that "extensive public participation in decision-making is essential for sustained development. Individuals, collectives and groups need to participate in the environmental impact assessment which may affect their decision-making; they should have the opportunity to obtain relevant information." "It is necessary to get important social groups to participate in policies and activities relating to all projects. NGOs play a crucial role in a participative democracy and enjoy a wealth of professional talents. The United Nations system and national governments should strengthen the mechanism of participation in decision-making by NGOs."

The core characteristic of modern governance is the diversification of governance subjects[17]—more attention should be paid to the roles of other social players including enterprises and civil social organizations while emphasizing the government's social management function. There is no exception for environmental governance. China is currently dedicated to building its national environmental governance capacity, while good governance is essential for it. Good governance of the environment needs to reflect many elements of the modern governance philosophy, including multi-subject participation, the improvement of the rule of law, the transparency of the decision-making—execution—management process, the effectiveness—efficiency—equity of decision-making and management, as well as accountability; more importantly, three mechanisms—the government's administrative management, market operations, and social supervision and interest check & balance—should be combined.[18]

---

16 Zhang Shiqiu, "Haze Control—Equal Importance Should Be Attached to Iron Hand and Good Governance," *Science and Society*, 2014 (2).
17 (France) Jean-Pierre Gaudin, *Pourquoi La Gouvernance*, trans. by Zhong Zhenyu. Beijing: Social Sciences Academic Press, 2010.
18 Zhang Shiqiu, "Reinforcing the Government's Environmental Responsibility Is Essential for Addressing Environmental Challenges," In *China Environmental Development Report (2013)*. Beijing: Social Sciences Academic Press, 2013.

As indicated by many surveys and much research, the central government and local governments act differently in such aspects as their developmental objective, and there is also local protectionism, so the government-led environmental governance and environmental policy cannot produce the desired effects. Bao Zhiming and Chen A'jiang believe that the current environmental struggle in China may be a governance dilemma, its root cause lies in the absence of a "society" rather than the specter of developmentalism.[19]

According to domestic and foreign experience, the development of an environmental civil society and NGOs helps not only to safeguard the environmental rights and interests of citizens, but it also improves the environmental social governance capacity and governance effect. In the opinion of Hong Dayong, as the environmental situation remains severe, only when China's environmental governance subjects and the environmental governance effect are improved by a rational and correct action of the safeguarding of the general public and public participation can the overall environmental situation be truly improved in an effective way.[20]

First, an environmental civil society and the NGOs for environmental protection are the important subjects in modern national environmental governance. According to the German politician Thomas Heberer, multi-subject participation is the important element in a modern governance system; it is very difficult for governments to handle all the affairs of a country, thus it is necessary to incorporate enterprises, society and citizens into public governance. This multi-subject governance mode is more effective than the single-subject government management mode.[21]

Second, environmental civil social organizations are the important forces for offering modern environmental public social services. The rapidly developing and increasingly active NGOs for environmental protection are the important lobbying and public benefit activity subjects for environmental protection and social supervision subjects, and are capable of providing effective environmental public services. Amidst a number of public affairs relating to environmental protection, the main suppliers of public services must extend to diverse subjects, including the NGOs, so as to make up for the deficiency of the government in management, control and services. The government can

---

19   Bao Zhiming and Chen A'jiang, "Environmental Dimension of China Experience: Dimension and Its Limit—Review and Rethinking of Sociological Study of China's Environment," *Sociological Study*, 2011 (6).
20   Hong Dayong, "A Tentative Study of the New Direction for Improving China's Environmental Governance," *Hunan Social Sciences*, 2008 (3).
21   (Germany) Thomas Heberer, "Promoting the Modernization of National Governance System and Governance Capacity," *People's Daily*, 2013-11-18.

purchase public services and adopt other means to gradually transfer to the third sector the public services, which cannot be addressed by the government and cannot be properly addressed by the government.[22] This diversifies the main suppliers of public services, reduces their costs and provides conditions for the development of social organizations and NGOs for environmental protection.

Third, environmental civil social organizations serve as an effective way of expanding orderly political participation in environmental protection by citizens, and they are conducive to avoiding accumulation of social contradictions caused by environmental problems by pushing forward rational participation and correct safeguard. First, with the participation in NGOs for environmental protection, citizens express their environmental appeals to the government, society and enterprises by means of the NGOs for environmental protection and so the channels for expressing individuals' environmental rights and interest appeals are expanded; moderate and rational activities regarding the safeguarding of environmental rights are carried out to protect the public environmental rights and interests of the citizens. Second, the NGOs for environmental protection serve as an important way for citizens to participate in managing national environmental affairs and the channels for citizens to take part in environmental politics and decision-making. It is particularly important that the NGOs for environmental protection are able to effectively guide the citizens to orderly participation in environmental public affairs according to laws so that a stable social counterweight is shaped and the increasingly acute conflicts of interest regarding environmental rights do not become extreme events of conflict, which would affect lasting social stability.

3 The Prospects of Development for China's Environmental Civil Society and NGOs for Environmental Protection

The society, the economy and the environment are closely related. Great progress has been made in China's environmental protection, but the environmental situation remains severe and the work on environmental protection faces even more severe challenges: (1) the environmental damage from accumulated and new environmental pollution and the corresponding public expenditures are still on the increase; (2) environmental risks still exist and are on the rise;

---

22   Li Huadong, "A Brief Analysis of the Roles of Civil Social Organizations in National Governance Modernization," *New West*, 2015 (2).

(3) environmental problems have and will continue to become the source of social conflicts; (4) the contradictions between the environment and development are acute, the conflicts of interest are intensifying, so the work on environmental protection still meets with an enormous number of difficulties. The potential for economic development is worrying. The principle causes mainly include the following: easier said than done in economic transformation; governance at the source remains highly difficult; the social and economic losses incurred by environmental pollution are huge; health risks are sustained; the long-term real effect of the improvement of social welfare amidst China's economic growth is questionable; the environmental capacity, ecosystem, natural resource endowment and public health guarantee are the bottlenecks in which breakthroughs must be made in order for there to be a stable economic and social development; in terms of weighing and balancing the short-term and long-term relationships among people's livelihood, development, ecology and the environment, the road remains long in theories, policies and measures; the current institutional costs for environmental management are high and efficiency is low, it is still difficult to favorably guide and influence economic transformation and structural adjustment; the mode of the dissemination of information is changing; the environmental awareness of the general public is improving; there are increasing appeals for safeguarding environmental rights and interests including the rights to know and to supervise, the right to participation and the right of claim; public pressure is growing—on the one hand, this provides great opportunities for environmental protection; on the other hand, the work on environmental protection may focus on short-term effects. As mentioned above, the development and maturation of an environmental civil society, the growth of NGOs for environmental protection are requisites for good governance of the environment and also for the accumulation of an important social capital for China's long-term sustainable development.

The newly revised *Environmental Protection Law* clarifies the higher legal status and requirements for public participation and the management, by the NGOs for environmental protection, of public benefit activities for environmental protection. In particular, the *Interpretation of the Supreme People's Court of Several Issues Concerning Applicable Laws for Hearing the Cases Involving Environmental Civil Public Interest Litigation*, issued on January 6, 2015, sets forth more specific provisions for the part concerning environmental public interest litigation in the *Environmental Protection Law*, officially effective as from January 1, 2015, and the revised *Civil Procedural Law*.

According to an estimation by the Ministry of Civil Affairs and this document, about 700 social organizations comply with relevant conditions and requirements and are qualified for filing environmental public interest litigations.

Under the new situation, besides public participation, information disclosure and confirmation of the legal status of the NGOs for environmental protection, the development of an environmental civil society and NGOs for environmental protection meets with many other important opportunities. First, the general public's awareness of the environment, environmental rights and interests, the safeguarding of rights, the participation in and enthusiasm for participation are increasing, thus providing the necessary social conditions for and laying the foundation for the development of an environmental civil society and NGOs for environmental protection; second, the structure of China's social governance has gradually changed from the centralization of power to the decentralization of power, offering more political opportunities and space for the development of an environmental civil society and NGOs for environmental protection; third, student organizations for environmental protection are rapidly developing, more academically professional young people are willing to work for the environmental public welfare undertakings, contributing to the enhancement of the professional competence of the environmental protection organizations; fourth, the development of modern information technology and the change in the mode of information dissemination make it impossible to monopolize information and also provide the most inexpensive and convenient conditions for the NGOs for environmental protection and for the general public to participate in environmental protection decision-making and for citizens to exercise supervision over the environment; fifth, the government is rapidly increasing the environmental investments and is starting to massively purchase public services from the NGOs for environmental protection; sixth, the attention from various foundations, enterprises and social capital towards the environmental public welfare undertaking has greatly increased, so there is a higher possibility that the NGOs for environmental protection, which were originally stagnant and hardly survived, will obtain the necessary fund support. However, challenges are greater amidst opportunities. These challenges exist in large quantities, the most crucial ones can be summarized as follows: first, much progress has been made in environmental protection law with respect to public participation and the legal status of the NGOs for environmental protection, but the development of the NGOs for environmental protection is still subject to many institutional restrictions; second, how can the NGOs for environmental protection can remain neutral or "politically correct" under legal authorization, public pressure and environmental interest conflicts? This tests the wisdom of the NGOs for environmental protection and the modern governance wisdom of the Chinese Government;

third, the causes for environmental problems are complicated, many subjects are affected, the social and economic impact is extensive, solutions are diverse and exert a widespread influence; therefore, besides great attraction, charisma and strong execution, the NGOs for environmental protection should also develop a higher level of expertise, so that their suggestions, appeals and participation can serve the specific purpose and be effective. The development and maturation of the NGOs for environmental protection helps ensure that citizens' environmental rights and interests are realized; most importantly, the government may try and push forward the change in China's social governance mode through the public issues and affairs—environmental protection, accumulating the necessary social capital for China's smooth transition to a harmonious society.

CHAPTER 2

Lessons from "APEC Blue" about Air Pollution Control in China

*Zhao Lijian**

Abstract

A plenty of measures were taken to control air pollution and thus improve the air quality in Beijing and its surrounding areas during the APEC Meeting in 2014. According to Beijing's official evaluation, as various measures were adopted during that meeting, the emission of pollutants was reduced by 50% on average, while the air quality was improved by 61.6% compared with that in the case of not taking those measures. The revelation from "APEC Blue" indicates that air pollution is preventable and controllable, but long-term improvement of the air quality cannot rely on short-term means and it should require long-term measures, while the transformation of the economic growth pattern and vigorous law enforcement supervision are particularly important. Air improvement entails costs, but the win-win outcome—clean air and economic growth—can be achieved. In order to turn "APEC Blue" into "Normal Blue" and "China Blue," China should carry out the following work: improve the legal and regulatory foundation and reinforce law enforcement supervision; establish a perfect policy framework and a management system; strengthen scientific decision-making and the executive capability of local governments; give full play to the roles of the economic and market means; further reinforce disclosure of environmental information and public participation.

Keywords

APEC Blue – air pollution control – environmental law enforcement supervision – cost-benefit analysis – policy evaluation – information disclosure – public participation

---

* Zhao Lijian, Director of China Environmental Management Project, Energy Foundation.

1 Origin of the Term "APEC Blue": from a Joke Told by the General Public in Response to a Political Commitment Made by Leaders

The leaders of 21 countries met in Beijing during the APEC Economic Leaders' Meeting hosted in Beijing, China during the period November 5–11, 2014. In order to guarantee good air quality in Beijing during this meeting, the *Air Quality Guarantee Plan for the APEC Meeting* was initiated and implemented in Beijing, Tianjin, Hebei, Shanxi, Inner Mongolia and Shandong as from November 3. Beijing, which had experienced haze many times in October, gladly welcomed the blue sky. The general public shared photos to hail the blue sky and called it the "APEC Blue." "APEC Blue" soon became a new term. Some people interpreted it as transient, unreal beauty. There was a typical sentence: "He does not like you so much, he is only APEC Blue"! Various jokes and explanations about "APEC Blue" were posted online; for example, "APEC Blue" was explained as "Air Pollution Eventually Controlled."

On November 10, Xi Jinping surprisingly spoke eloquently of "APEC Blue" when making a speech before the APEC welcoming banquet. He said,

> During these days, the first thing I have done after getting up every morning is that I have checked the air quality in Beijing; I hope that the haze can be mitigated somewhat so that our guests from afar feel a bit comfortable in Beijing.... Some people have said that Beijing's blue sky is 'APEC Blue', beautiful but short-lived. I hope and believe that, with unremitting efforts, 'APEC Blue' can be maintained.... I hope that a blue sky, green hills and clear waters will always be available in Beijing, and even nationwide, and that our children will be able to live in a good ecological environment. This is a very important part of the Chinese Dream.[1]

Therefore, "APEC Blue" became a political commitment from a joke among the ordinary people; it reflects the confidence and political wisdom of the Chinese leader. According to the speech delivered by Xi Jinping, the Chinese Government does not control air pollution merely for the purpose of the APEC Meeting. It is not a short-term act; instead, it is a long-term commitment. The Chinese Government does not control air pollution only for the capital city of Beijing; the Chinese Government is determined to control air pollution all across China. Such a commitment, especially a commitment made before leaders of many countries around the world, is a weighty one.

---

1 Delivering a Speech at the APEC Welcoming Banquet. www.cyol.net, http://news.xinhuanet.com/world/2014-11/11319112.htm.

Such an open declaration and commitment from the top leader will greatly push forward air pollution prevention and control and environmental protection in China because much attention as well as a commitment from the government is the precondition for solving public problems such as air pollution. This attention should not merely come from the government's environmental protection department as the problems concerning air pollution, even environmental pollution, always involve various aspects of the social and economic life, including economic development, the energy structure, the industrial structure and transportation. Therefore, people merely count on the government's environmental protection department, since many hard measures involving macro decision-making and coordination among departments are very difficult to be introduced, or it is difficult to implement them even if they are introduced.

The attention from the government is reflected in various ways; one of these ways is that the State Council unveils relevant documents or action plans at the central level, while the provincial and municipal governments introduce relevant documents or action plans at the local level. The *Action Plan for Air Pollution Prevention and Control* released by the State Council in September 2013 and the subsequent action plans adopted in provinces and municipalities constitute an example. Another indicator for checking whether the government pays attention to air pollution control is the frequency of discussing the work on air pollution control in the executive meetings of the State Council or provincial and municipal governments. In the past 2–3 years, the frequency of specially discussing or focusing discussions on air pollution control in these executive meetings has significantly increased, showing that the degree of attention from the government has risen greatly.

2   Measures for "APEC Blue" and Their Effects

The work on guaranteeing a good quality of air for APEC Meeting was arranged as early as 2013. The Ministry of Environmental Protection and the authorities of Beijing and the surrounding five provinces, autonomous regions and municipalities jointly studied and prepared the *Air Quality Guarantee Plan for the APEC Meeting* ("Guarantee Plan"). This plan was deliberated and adopted in the third meeting of the group for coordinated prevention and control of air pollution in the Beijing-Tianjin-Hebei Region and the surrounding areas as chaired by Vice-premier Zhang Gaoli in October 2014. According to relevant information from several experts, this plan is basically similar to the air quality guarantee plan carried out during the Beijing Olympic Games in 2008; however, prevention and control efforts were greatly intensified because of adverse meteorological conditions during the practical implementation of the plan. As shown by the preliminarily estimated data from the Ministry of Environmental Protection, 9,298 enterprises actually halted production, 3,900 enterprises limited production and more than 40,000 construction sites were shut down in six provinces, autonomous regions and municipalities during the APEC Meeting, being 3.6 times, 2.1 times and 7.6 times that specified in this plan.[2]

What were the effects produced by these measures? What was the extent to which the quality of the air was improved in terms of data, besides the visual "APEC Blue" effect? On December 17, 2014, Beijing Municipal Environmental Protection Bureau released the result of the evaluation of the effect of the air quality guarantee measures during the APEC Meeting. According to a relevant report, the average PM2.5 concentration was 43 μg/m³ in Beijing during the APEC Meeting; if no guarantee measures had been jointly adopted in Beijing and its surrounding areas during the APEC Meeting, PM2.5 concentration was expected to reach 69.5 μg/m³ during that Meeting, 61.6% higher than the actual concentration. In this evaluation, the contributions from measures adopted in Beijing were also analyzed[3] (see Table 2.1).

TABLE 2.1   Air pollution control measures adopted in Beijing during the APEC Meeting and their contributions to emission reduction

Unit: %

Beijing's measures	Contributions to pollution emission reduction
Imposition of restrictions on the use of motor vehicles on roads	39.5
Shutdown of construction sites	19.9
Production halt and limitation on work in enterprises	17.5
Arrangement of holidays and adjustment of rest days for citizens	12.4
Road cleaning	10.7

SOURCE: BEIJING MUNICIPAL ENVIRONMENTAL PROTECTION BUREAU

2   Regional Interaction, Multiple Measures and Careful Deployment Fully Guarantee the Air Quality during the APEC Meeting. Website of the Ministry of Environmental Protection of the People's Republic of China, http://www.mep.gov.cn/gkml/hbb/qt/201411/t20141115_291482.htm.
3   Beijing Releases the Results of the Evaluation of the Effect of the Air Quality Guarantee Measures during the APEC Meeting. Website of Beijing Municipal Environmental Protection Bureau, http://www.bjepb.gov.cn/bjepb/324122/416697/index.html.

3   Revelations from "APEC Blue"

We can obtain the following revelations from the air quality pollution control measures adopted during the APEC Meeting and their effects.

Revelation One: Air pollution is preventable and controllable. The adoption of measures during the APEC Meeting looked like a large experiment, in which various means were taken to reduce the emission of pollutants; as a result, pollutants were decreased by about 50% on average, significantly improving the quality of the air. The relationship between the man-made reduction of pollutant emissions and the substantial improvement in the quality of the air has been recognized extensively; in particular, it has enhanced the confidence of government departments: as long as great efforts are made to reduce the emission of pollutants, the quality of the air can be improved.

Revelation Two: The improvement in the quality of the air during the APEC Meeting mainly relied on short-term measures; however, the long-term improvement in the quality of the air depends heavily on long-term measures. The measures adopted during the APEC Meeting were basically such short-term measures as an imposition of restrictions on the use of motor vehicles on roads, the shutdown of construction sites, a production halt and limitations on work in enterprises, the arrangement of holidays and the adjustment of rest days for citizens, an intensification of road cleaning. Most of these short-term measures cannot become normal, while long-term improvement in the quality of the air requires the adoption of long-term measures, including strengthening legislation and law enforcement, tightening the pollution emission standards, improving the energy structure and the industrial structure as well as the system of public transport and non-motor vehicle driving. However, in case an early warning of heavy pollution occurs in the future, short-term measures may be adopted again even though they will cause many inconveniences to industrial production and the life of citizens. We should choose the lesser evil among production, inconveniences in everyday living and a huge impact on our health—we should endure short-term inconveniences in everyday living to mitigate the damage to our health. Moreover, a halt in industrial production and inconveniences in everyday living will constantly remind the government and the general public of resolutely adopting more long-term measures to reduce air pollution, thus decreasing and avoiding these inconveniences. In this sense, the short-term measures for bringing the "APEC Blue" back can also, and should, be implemented again under the particular circumstances—in case of an early warning of heavy pollution. In this way, the health of the general public is protected and the government can more calmly confront the criticism that "the government only cares about face and comfort for foreign officials, but it pays no attention to the health of the general public."

Revelation Three: Reinforcing Law enforcement supervision are both short-term and long-term effects; violations of environmental law still exist extensively in China. An analysis of the effects from the measures of creating "APEC Blue" in Beijing does not cover the effects from strengthening law enforcement supervision. However, obviously, the Ministry of Environmental Protection and the governments of Beijing and the surrounding provinces, autonomous regions and municipalities intensified law enforcement supervision, which substantially reduced the emission of pollutants beyond relevant standards, the surreptitious emission of pollutants and leak-induced emission of pollutants from industrial enterprises during the APEC Meeting and considerably lowered pollutant emissions. This can partly explain the reason why pollutants were decreased by 50% on average during the APEC Meeting and the quality of the air was improved by 61.6% compared with that in the case of not taking measures. Generally, the improvement in the quality of the air is smaller than the reduction of the amount of pollutant emissions.

Meanwhile, according to the analysis of law enforcement inspection during the APEC Meeting, it is absolutely necessary to improve the supervision of environmental law enforcement in China! In order to guarantee good quality of the air during the APEC Meeting, the Ministry of Environmental Protection organized 16 supervision and inspection groups to carry out supervision and inspection as from November 3, 2014. They conducted overt observations and secret investigations, inspected the implementation of the measures specified in the *Air Quality Guarantee Plan for the APEC Meeting* in different areas. According to the data from the Ministry of Environmental Protection, as of the wee hours of the 5th day, these 16 supervision and inspection groups had inspected 395 enterprises: 317 enterprises were included in the list of a halt in and limitation to production, among which 33 enterprises had not halted and limited their production according to the requirements; 78 enterprises were not included in the list of a halt in and limitation to production, among which 31 enterprises caused pollution emission beyond relevant standards and 25 enterprises showed a failure to ensure the normal operation of air pollutant control facilities. They inspected 103 construction sites, among which 18 construction sites did not stop construction according to requirements and 37 construction sites indicated to have an insufficient implementation of flying dust control measures. Furthermore, they found 110 sites of straw garbage incineration, 30 road sections where the flying dust problem was severe and 37 other atmospheric environmental problems.[4] It should be specially noted

---

4 "Air Quality Guarantee Measures Were Not Duly Implemented in Some Areas." Website of the Ministry of Environmental Protection of the People's Republic of China, http://www.mep.gov.cn/gkml/hbb/qt/201411/t20141106_291198.htm.

that, among 78 inspected enterprises not included in the list of a halt in and limitation to production, 31 enterprises emitted pollutants beyond relevant standards and 25 enterprises failed to ensure the normal operation of air pollutant control facilities; in other words, 71.8% of the enterprises had violated the environmental law! Law-breaking enterprises made up an extremely high proportion of intensified environmental law enforcement during the APEC Meeting, let alone during ordinary times.

Revelation Four: Air pollution control should comprehensively consider costs and benefits so as to achieve a win-win outcome—clean air and economic growth. According to *The Wall Street Journal*, as estimated by Credit Suisse, the short-term measures adopted in six provinces, autonomous regions and municipalities during the APEC Meeting affected about 1/4 of the steel production, 13% of the cement production and 3% of the industrial output value in China. In other words, as a consequence, China's industrial output value decreased by 0.2%–0.4% in November.[5] This had a great impact on China's economy, which was slowing down. Besides these short-term measures, the long-term measures for controlling air pollution also entail costs; for example, the addition of pollution end control technology will increase costs for enterprises and for their products; the improvement of the energy structure and urban transportation needs massive public investments; closing down outdated production facilities will incur economic losses to enterprises and will cause unemployment and other social problems.

Air pollution control measures involve costs, but inaction in controlling air pollution also entails costs. Owning clean air has become the important prerequisite for a powerful economy. Clean air can reduce public health costs, increase crop output, mitigate damage to raw materials and infrastructure and lower the costs for treatment after pollution. Clear air is also an important factor for attracting and retaining top business talents around the world to work and live in the country.

This once again reminds us that any public policy should weigh advantages and disadvantages on various aspects and we cannot adopt ideal measures for a single objective, so the cost-benefit analysis of policies and measures is particularly important. At the present stage, China has no well-enhanced mechanism for making policies and measures based on a cost-benefit analysis. When developing and evaluating the *Clean Air Act*, the USA analyzed its costs and benefits and found that, in 20 years from 1970 to 1990, 1 USD cost

---

5   Mark Magnier et al., "APEC Blue Has Chinese Factories Bleeding Red, Economists Say." *Wall Street Journal*, 2014-11-12, http://blogs.wsj.com/chinarealtime/2014/11/12/apec-blue-has-chinese-factories-bleeding-red-economists-say/.

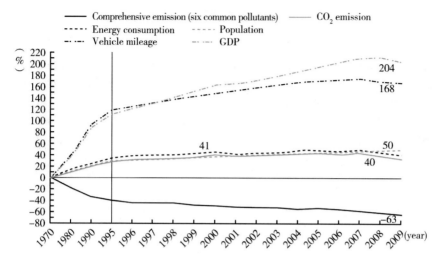

FIGURE 2.1 Comparison of the trend of air pollutant emissions and that of economic and social development in the USA, 1970–2009

for controlling air pollution brought a health benefit worth 40 USD; in the 20 years from 1990 to 2010, that figure was 4 USD; in general, the measures adopted for controlling air pollution generated positive benefits to the whole society. The result of air pollution control in the USA also reflected positive benefits; as shown in Figure 2.1, from 1970 to 2009, the emission of pollutants in the USA generally decreased by 63%, while the USA's GDP increased by 204%.[6,7] A win-win outcome—air pollution control and economic growth—was achieved.

Other important developed countries and territories in the world such as Europe and Japan also experienced severe air pollution; now they have basically gotten air pollution under control, while the control of air pollution was not an important factor affecting their overall economic performance.

Therefore, air pollution control measures incurred costs and may affect economic growth in the short term, but in the long term, the win-win outcome—clean air and economic growth—can be achieved. This has been proved in many countries, and China can also achieve this outcome.

---

6 EPA, "The Benefits and Costs of the Clean Air Act, 1970 to 1990," 1997, http://www.epa.gov/cleanairactbenefits/retro.html.
7 EPA, "The Benefits and Costs of the Clean Air Act, 1990 to 2010," 1999, http://www.epa.gov/cleanairactbenefits/prospective1.html.

4 How to Turn "APEC Blue" into "Normal Blue" and "China Blue"

The government and society have reached a consensus that "APEC Blue" should become the "Normal Blue" which is not limited to a certain period of time and "China Blue" which goes beyond a certain area and covers the whole of China. However, this objective cannot be achieved overnight. Even in the case of "APEC Blue" achieved by vigorously controlling air pollution, the average PM2.5 concentration during the APEC Meeting was 43 μg/m^3, higher than the national standard of 35 μg/m^3.

Take a look at the improvement in the quality of the air in 2014; much progress was made, but the task remains heavy and the road is still long. According to the air quality data from the Ministry of Environmental Protection with respect to 74 cities,[8] including the Beijing-Tianjin-Hebei Region, the Yangtze River Delta, the Pearl River Delta, municipalities directly under the Central Government, provincial capitals and cities specifically designated in the state plan in 2014, only 8 cites reached the national grade II standard of 35 μg/m^3. This represented progress compared with 2013 when only three cities, Haikou, Lhasa and Zhoushan, reached the standard. In particular, the cities which reached the standard included Shenzhen, an economically developed city with a population of 10 million, so there was a certain significance of a benchmark. The PM2.5 concentration in Beijing in 2014 declined by 4% compared with 2013, below the annual target of 5%. The PM2.5 concentration in the other areas of the Beijing-Tianjin-Hebei Region, the Yangtze River Delta and the Pearl River Delta decreased by more than 10%; the monitoring data showed that the quality of the air generally improved compared with the previous year.

According to the data from the Ministry of Environmental Protection, among 10 cities with a relatively low quality of air in 2014, 7 cities in Hebei Province were involved, which were Baoding, Xingtai, Shijiazhuang, Tangshan, Handan, Hengshui and Langfang, while the remaining cities were Ji'nan, Zhengzhou and Tianjin. However, this did not mean that the quality of the air in Henan or Shandong was better than that in Hebei because Hebei falls within a major area of the Beijing-Tianjin-Hebei Region and the provincial capital, the cities specifically designated in the state plan, and all of the other prefecture-level cities in Hebei were included in these 74 cities under key monitoring, while only the provincial capitals and the cities specifically designated in the state plan in Henan and Shandong were included in those cities.

---

8 The Ministry of Environmental Protection released data concerning the quality of the air in major areas and 74 cities in 2014. Website of the Ministry of Environmental Protection of the People's Republic of China, http://www.zhb.gov.cn/gkml/hbb/qt/201502/t20150202_295333.htm.

In 2014, long-lasting extensive heavy pollution weather occurred twice (in February and October) nationwide, and the frequent heavy pollution weather did not fundamentally improve. Various areas were still frequented by heavy pollution. The Beijing Marathon and the Tianjin Marathon were shrouded under a heavy blanket of air pollution, arousing public attention.

Many things should be done to really turn "APEC Blue" into "Normal Blue" and "China Blue."

4.1 Improve the Legal and Regulatory Foundation and Strengthen Law Enforcement Supervision

In December 2014, the draft amendment to the *Air Pollution Prevention and Control Law* submitted by the State Council was deliberated in the Standing Committee of the National People's Congress, so the law which had not been amended in 15 years entered the channel for accelerated amendment. In August 2015, the *Air Pollution Prevention and Control Law*, newly amended by relevant departments, was ultimately adopted. The people have paid great attention to air pollution in the past two years, so the amendment to this law has also attracted a lot of attention. A vigorous law will lay a solid legal foundation for controlling China's air pollution in the next 10–20 years. Several key points of national legislation can be described as follows: How to ensure a mechanism for continuously amending the air quality standard and providing a long-term guarantee for public health; how to build a mechanism for managing compliance with standards and specify stricter requirements for the areas below standards; how to make a pollutant emission license become the effective means of law enforcement supervision; how to intensify punishment of violations of laws; how to ensure the effective means for the disclosure of environmental information and public participation. Meanwhile, local legislation concerning air pollution prevention and control seemed to forestall legislation at the central level—*Beijing Air Pollution Prevention and Control Regulations* entered into force in Beijing as of March 2014, and local legislation involving air pollution prevention and control has also been carried out in Shanghai, Tianjin and Jiangsu.

With regard to law enforcement supervision, the Ministry of Environmental Protection reported supervision and inspection of air pollution prevention and control in December 2014.[9] It hoped that such work could be maintained as normal in the future and has required local authorities to carry out such work.

---

9 The Ministry of Environmental Protection reported supervision and inspection of air pollution prevention and control in December, 2014. Website of the Ministry of Environmental Protection of the People's Republic of China, http://www.mep.gov.cn/gkml/hbb/qt/201502/t20150204_295454.htm.

4.2 Build a Sound Policy Framework and Management System

Much of the international experience shows that sustained improvement in the quality of the air is not a simple task; it cannot be achieved overnight and calls for tangible efforts and integrated solutions. Short-term actions are not enough. It is essential to establish an excellent policy framework and an exquisite management system. Such a policy framework and management system should be based on some basic principles, including: (1) study and build a sophisticated air quality management structure; (2) ensure the input of adequate personnel and funds; (3) make decisions as a result of the latest scientific analysis; (4) establish a heavy pollution emergency early warning and response system; (5) develop air pollution prevention and control measures and choose the preferred measures on the basis of a cost-benefit analysis; (6) apply the best feasible technologies; (7) make coordinated efforts to control air pollutants and greenhouse gases; (8) adopt incentive and punishment measures to ensure the implementation and the enforcement of the laws; (9) strengthen information disclosure and encourage public participation; (10) carry out regular monitoring and evaluation to achieve sustained improvement.

4.3 Reinforce Scientific Decision-Making and the Local Governments' Capability for Implementation

The policy evaluation report *Whether the PM2.5 Improvement Goal 2017 Can Be Achieved in the Beijing-Tianjin-Hebei Region* jointly released by the Clean Air Alliance of China (CAAC) and Tsinghua University in September 2014 has drawn wide attention. This scientific evaluation of policies and measures was not conducted by local governments in developing and executing local air pollution prevention and control action plans. The government should consciously and willingly make efforts in this regard; scientific research institutions and social organizations should work more closely with the government; more policy tools, relevant training and capacity building are required. In 2014, air pollution prevention and control training among mayors from the Beijing-Tianjin-Hebei region and special air pollution prevention and control training among directors of environmental protection bureaus nationwide were conducted. This kind of professional training should be conducted in a more widespread fashion.

4.4 Give Full Play to the Roles of the Economy and the Market

In 2014, the pollutant charge standard increased by 7 times and 10–20 times in Tianjin and Beijing, respectively. These measures were designed to push ahead with air pollution prevention and control by means of the economy and the market. In September 2014, the National Development and Reform Commission, the Ministry of Finance and the Ministry of Environmental Protection issued a *Circular Concerning Issues Involving the Adjustment of the Pollutant Charge Standard*, which increased the pollutant charge standard at the national level. The examples in Beijing and Tianjin tell us that the previous pollutant charge standard was too low, while more local authorities should take action as soon as possible.

4.5 *Strengthen Information Disclosure and Public Participation*
Great progress was made in information disclosure and public participation in the information field in 2014. Since then the government and enterprises have advanced in information disclosure. In particular, the amendment to the *Environmental Protection Law* opened the door to future environmental information disclosure and public interest litigation. The Institute of Public & Environmental Affairs (IPE), the Friends of Nature and a number of organizations have made positive progress in pollution information disclosure through joint actions. Environmental protection APPs launched in 2014 collect enterprise pollution emission information in a unified way, making such information visible, while the well-targeted publicity carried out among the enterprises with emissions exceeding relevant standards and imposition of pressure on them on the basis of this information have become increasingly effective. The *Listed Companies amid Haze* has exerted pressure on listed companies violating environmental law and emitting pollutants beyond relevant standards through public investors. SEE Foundation, Alibaba Foundation, and Weilan Foundation supported by Energy Foundation support public interest organizations in various areas to carry out surveys and public interest activities relating to air pollution prevention and control. Greenpeace analyzes air quality information and develops a ranking list. The World Wide Fund for Nature (WWF), the Energy Foundation and the Clean Air Alliance of China carry out the "Earth Hour: Blue Sky Self-made" activity. The "Blue Sky Reappears" public service advertisement about wildlife rescue is presented by stars. All of these initiatives have, to a large extent, played the active role of environmental protection organizations and have aroused public enthusiasm for participation. Finally, air pollution in China cannot be eliminated without extensive public participation; with the adoption of a better mechanism for mass prevention and mass control, "APEC Blue" is expected to be soon turned into "Normal Blue" and "China Blue."

CHAPTER 3

Controlling Aggregate Coal Consumption: Plans and Policies

*Chen Dan, Lin Mingche, and Yang Fuqiang**

Abstract

This chapter stresses that the total quantity control of coal consumption is crucial for the low-carbon green transformation of energy in China. Coal mining, transportation and its utilization exert a great impact on ecology, the environment, public health and climate change, and incur very high external costs in social and economic fields. Coal is the chief culprit for air pollution, haze, solid wastes and water pollution. China is the largest carbon dioxide emitter. Coal consumption must first be reduced prior to reaching its peak by 2030.

Keywords

total quantity control of coal consumption – haze – public health – climate change

China is the largest energy producer and consumer in the world. In 2014, the total energy consumption hit 4.26 billion tce and coal accounted for 66% of the total energy consumption.[1] In the same period, coal made up 29.9% of the structure of the world's energy consumption, and it accounted for about 20% of that in the OECD countries.[2] The proportion of carbon dioxide emissions

---

* Chen Dan, former senior analyst of the issue concerning total quantity control of coal under the China Climate and Energy Program of the Natural Resource Defense Council (NRDC), has taken charge of and coordinated the research on about 20 issues. Lin Mingche, Director of the China Climate and Energy Policy at NRDC, resident at NRDC's Beijing Office, has focused on analyzing China's climate and clean energy policy, including how China expands its energy efficiency and renewable energy resources, and has provided incentive measures and built systems for low-carbon development. Yang Fuqiang, PhD, NRDC's senior advisor in climate change, energy and environment, has engaged in researching China's sustainable energy strategy and policy in nearly 40 years.

1 National Bureau of Statistics, *China Statistical Yearbook 2013*. 2014-02-25.
2 *BP World Energy Statistics 2014*. BP, 2014-06.

from global coal consumption in energy activities has stood at about 40% for a long time, much lower than that in China, which has been about 80%. Carbon dioxide emissions from global coal consumption have risen since 2000, which has mainly been due to the great increase in China's consumption of coal.

Since the reform and opening up, remarkable achievements have been made in China's economic development. In 2012, China became the second largest economy in the world. The coal-dominated energy supply has made China's economy prosper, but it has also cost China a heavy price—the environmental and ecological system has been severely damaged and public health has been subject to threat. At present, this trend of deterioration has not yet been fundamentally contained. China's resources are being rapidly consumed; in particular, land and water resources have become more and more scarce at the regional level due to a type of development marked by the depletion of mineral resources. China is the largest carbon dioxide emitter in the world. Given the current developmental trend, carbon dioxide emissions in China will account for about 28% of the total carbon dioxide emissions in the world by 2020.[3]

1   The Air Pollution Crisis Gave Rise to the Total Quantity Control of Coal Consumption

The frequent extensive heavy haze nationwide in recent years has sounded the alarm. China is now at the critical juncture of having to resolutely address the unbalanced, uncoordinated and unsustainable development among the economy, resources and the environment. In September 2013, the State Council released the *Action Plan for Air Pollution Prevention and Control*.[4] This action plan explicitly states that China must set the national medium and long-term control objectives for the total quantity of coal consumption and exercise objective responsibility management. In 2014, coal reduction, replacement and cleaning were the primary effective measures for coping with the deterioration of the quality of air and reducing the PM2.5 concentration. Currently, the authorities in many eastern coastal provinces have specified the control objectives for the total quantity of coal consumption; for example, the authorities

---

3 Lin Mingche, Yang Fuqiang, et al., "China's Sustainable Energy Development Strategy," *China Energy*, 2013, (1–2).
4 General Office of the State Council, "The State Council Releases Ten Measures in the *Action Plan for Air Pollution Prevention and Control*." http://www.gov.cn/jrzg/2013-09/12/content_2486918.htm.

in Beijing, Tianjin, Hebei, Shandong, the Yangtze River Delta, the Pearl River Delta, and Shaanxi Province in the western China have developed the plans for the total quantity control of coal. Haze pollution has spread across China. The city groups in central Liaoning Province, Shandong Province, the Guanzhong region of Shaanxi Province and the Gansu-Ningxia region, the Wuhan city group, the Changsha-Zhuzhou-Xiangtan city group, the Chengdu-Chongqing city group, and the Urumqi city group of Xinjiang have been hit by haze, while the city group on the west coast of the Taiwan Straits has also experienced air pollution in a rapidly increasing number of days. With urbanization and economic expansion, air pollution will continue to worsen nationwide. Among 74 cities under the PM2.5 monitoring in 2013, only 3 cities met the requirement of the annual average concentration of 35 μg/m^3 (the first phase standard of the World Health Organization). In order to cope with air pollution, this action plan states that the proportion of external power transmission should be gradually increased, the natural gas supply and the intensity of non-fossil energy utilization should be raised and other measures should be adopted to replace coal as a fuel.[5] These measures have drawn public attention. At present, not a few new coal-fired power plant projects have been approved to be carried out in the main national coal bases. Moreover, some coal-to-gas projects have also been granted permits. The production capacity is expected to reach 50 billion m^3 by 2020. These measures have raised concerns about pollution transfer and an increase in carbon dioxide emissions. To effectively control air pollution nationwide, coal utilization should be reduced in the Beijing-Tianjin-Hebei Region, in the Yangtze River Delta and in the Pearl River Delta, and coal consumption should also be controlled at the same time in other regions. The authorities in regions, provinces and municipalities should follow the basic principle that "the air quality cannot deteriorate" and develop the local environmental capacity indicators. Therefore, it is imperative to set and implement the goal of the total quantity control of coal consumption nationwide, in various regions and main coal-consuming sectors.

---

5 General Office of the State Council, "The State Council Releases Ten Measures in the *Action Plan for Air Pollution Prevention and Control.*" http://www.gov.cn/jrzg/2013-9/12/content_2486918.htm.

2   China is Vigorously Pushing Forward Its Low-carbon Transformation

The Beijing Energy System Center under the Energy Research Institute of the National Development and Reform Commission has taken the perspective of looking forward and looking back and set three different scenarios for economic, social and energy development, including an energy-saving benchmark scenario, a coal consumption control scenario and a radical scenario.[6]

In the energy-saving benchmark scenario, coal consumption will steadily increase and the peak will not exceed 3.44 billion tce prior to 2030; after 2030, the substitution effect of natural gas, renewable energy and other types of energy will be enhanced and coal consumption will slowly decline, and it will decrease to 3.04 billion tce by 2050, but its proportion will be as high as 46.3%, and coal will continue to dominate energy consumption by 2050. The proportion of non-fossil energy will rapidly climb and reach 14% by 2020. The carbon emission intensity per unit of GDP by 2020 will decrease by 45% compared with 2005.

In the coal consumption control scenario, coal consumption will reach the peak of 2.9 billion tce before 2020. Thanks to the rapid increase in natural gas and non-fossil electric power, coal consumption will gradually fall as from 2020 and will be only 1.68 billion tce and its proportion will dramatically decrease to 28.8% by 2050. The proportion of non-fossil energy will rapidly rise to 15%, 22% and 40% by 2020, 2030 and 2050 respectively, and non-fossil energy will surpass coal to become the dominant energy. Carbon emissions will reach a peak by 2025 and will continuously, markedly decrease after 2030. The carbon emission intensity per unit of GDP by 2020 will decline by 48% compared with 2005.

In the radical scenario, coal consumption will reach the peak of 2.9 billion tce before 2018, while it will rapidly drop after 2018; it will decrease to 1.42 billion tce and will account for less than 26.5% by 2050. The proportion of non-fossil energy consumption will rapidly go up and hit 46% by 2050. Carbon emissions will reach a peak prior to 2022 and then rapidly descend, and it will decrease to the level of 2010 by 2050. The carbon emission intensity per unit of GDP by 2020 will decline by 50% compared with 2005.

---

6   The Research Group of the Energy Research Institute for "A Study of the Plan and Policy for Total Quantity Control of Coal Consumption in China." *A Study of Scenarios Involving the Total Quantity Control of Coal Consumption in China.* 2015-01.

Pushing forward the transformation of the energy structure has become an urgent task for China at the present stage. In order to achieve this goal, China should work on the following aspects.

2.1   Reform the Existing Energy Management Mechanism

The Chinese Government will adopt a plan for the total quantity control of carbon emission reduction in order to restrict the growth of carbon emissions in the 13th Five-Year Plan period from 2016 to 2020, and lay a foundation for establishing a national carbon trading market across China from 2016 to 2020. Currently, the *Renewable Energy Law*, the *Energy Conservation Law*, and the *Environmental Law* that has come into force in China have laid a very good legal foundation for energy conservation and emission reduction in China. However, strong legal support for climate change and carbon reduction is still not available in China. In pressing ahead with the national rule of law reform, relevant departments are actively promoting the making of the *Climate Change Promotion Law*. Furthermore, a support law—the *Energy Law*—has been introduced, and such laws as the *Air Pollution Prevention and Control Law* and the *Coal Law* have been revised. With the support of these laws, the intensity of carbon emissions and the total coal quantity can be controlled in terms of legislation and law enforcement procedures, laying a solid foundation for an absolute quantity reduction of carbon emissions.

2.2   Fighting against Environmental Pollution is an Important Force for Driving Forward the Reduction of Carbon Emissions

The ecological red lines and hard rules for environmental quality can stimulate enterprises to strictly observe environmental standards and actively reduce pollutant emissions. For enterprises, technologies and measures for cutting down coal consumption and more effectively reducing pollutant emissions can promote carbon reduction, and *vice versa*. As the environmental quality is improved, priority will be given to some projects with a good synergistic effect from investments, technologies and measures. Where appropriate, carbon dioxide should be classified as a pollutant and receive equal attention like other pollutants in emission control.

2.3   China is Establishing a Carbon Trading Market

The Chinese Government is popularizing the pilot carbon trading market in two provinces and five cities with the hope of establishing a national carbon market as soon as possible. Utilizing the carbon trading market mechanism is an important measure for China to actively address climate change. Given the stagnation, a number of congenital or acquired defects and weaknesses in the current international carbon trading market, China should avoid the problems of other countries in establishing the carbon market so that the Chinese carbon market can healthily develop and grow. Its preconditions include not only laws, but also the establishment of a data reliability system and a more effective combination of integrity and punishment measures.

2.4 Introduce a Resource Tax, a Carbon Tax and an Ecological Tax, and Cancel Fossil Energy Subsidies

It is necessary to actively adopt market levers such as taxation to reduce carbon dioxide emissions. A resource tax should be gradually introduced in the development of coal and other fossil fuels. A carbon tax is an excellent tax for reducing carbon emissions. At present, attention is given to how to closely combine a carbon tax with the carbon trading market rather than letting them oppose each other. Foreign experience tells us that a carbon tax and the carbon market can jointly come into play. The ecological tax is an effective means for differentiating carbon dioxide emission policies in different regions of China such as the eastern, central and western regions. The government subsidies for mineral energy should be cancelled so that energy prices are determined by market supply and demand, and so natural gas and other unconventional gases can be more rapidly developed and utilized. In the coal, oil and electric power sectors, subsidies should also be cancelled, and external costs should be gradually internalized and based on prices as well.

3 The Total Quantity Control of Coal Consumption Has Triggered a New Round of the Reform of the Energy Sector

3.1 The Plan for the Total Quantity Control of Coal Consumption Can Prove and Draw on International Experience

In the periods of the 1950s–1960s, energy consumption was dominated by coal in the developed countries, causing many environmental and public health problems; for example, the London smog incident occurred in December 1952, and thousands of people died due to air pollution in a week. At that time, many large cities in Europe and the USA were also shrouded by coal smoke which severely threatened public health and economic development. Environmental deterioration and city air pollution also aroused numerous public movements for environmental protection in the 1970s. The developed countries used plenty of available oil and natural gas to replace coal. Since then, the world has entered the stage of oil consumption from the era of coal consumption; coal has no longer been the main energy and has become secondary. The total coal consumption in the developed countries declined and then it became relatively steady. In recent years, against the general background of addressing climate change, total coal consumption has tended to further decrease in most of the developed countries. According to the experience gained from the developed countries, the reason for the decrease in coal lies not only in the market mechanism, but also in the appearance of alternative energy such as oil and natural gas; more importantly, coal consumption has exerted a severe impact on the environment and public health. In general, public health, environmental protection and the development of alternative energy are the three main forces for driving forward the reduction of coal consumption.

3.2 The Total Quantity Control of Coal Consumption is the Focal Point in the Total Quantity Control of Energy

The 12th Five-Year Plan specifies the goal of energy intensity and the plan for the total quantity control of energy consumption.[7] Both goals supplement each other. Coal is the central part in the total quantity control of energy consumption. Coal pollutants are generally 1.2–1.5 times those from natural gas, and carbon dioxide from coal is 1.7 times that from natural gas. The progress in total quantity control of energy consumption has been barely satisfactory in the past 1–2 years, which is mainly due to a failure to grapple with the central problems. On the one hand, we should control the total coal quantity and reduce coal consumption; on the other hand, we should stimulate the development of renewable energy and clean energy, and increase their proportion in total energy consumption. Only when we place emphasis on both aspects can the energy intensity be effectively reduced and can total energy consumption be controlled. At present, it is essential to focus on studying the plan and policy for the total quantity control of coal.

3.3 The Plan for the Total Quantity Control of Coal Consumption Can Speed Up Energy Transformation

China must transform its current pattern of economic development and its economic structure. The previous pattern with heavy pollution, high emissions, high input, low efficiency and low output has come to an end. The Central Government stresses reform and a five-in-one developmental path covering economic construction, political construction, cultural construction, social construction and the construction of an ecological civilization; moreover, it puts the construction of an ecological civilization in an important

---

7 National Bureau of Statistics of the People's Republic of China, *Statistical Communiqué of the National Economic and Social Development 2013*. 2014-02-24.

position, laying a solid foundation for sustainable economic development. Energy serves as the engine for economic development, while economic transformation cannot be achieved if the energy sector is not transformed in a clean way. China's current coal-dominated energy structure is high-carbon and inefficient. According to a preliminary estimate from the National Bureau of Statistics, in 2014, the total energy consumption was 4.26 billion tce and coal accounted for 66% of the entire energy consumption.[8] If energy transformation is not conducted, it will be very difficult to transform the pattern of economic growth and its structure. The key to energy transformation consists in reducing the dependence on coal consumption. A decrease in the total coal consumption across China in 2014 occurred for the first time since the reform and opening up that began in 1978, so it is of profound significance.

4 The Total Quantity Control of Coal Consumption Promotes the Realization of the Goals for Environmental Protection, Resource Conservation and Climate Change

4.1 *The Total Quantity Control of Coal Consumption Propels a Sustainable Development of the Coal Industry*

As shown by many studies, at present, coal production and development are immensely ruining the environment and resources, and the production capacity which is green, safe and efficient accounts for only 40%–60% of the coal output.[9] According to relevant statistical data, in 2012, 560 million tons of coal gangue were generated, 6.2 billion tons were stockpiled nationwide, covering an area of 20,000 hectares. Autoxidation and slow oxidation of gangue heaps have discharged about 1.10 million tons of sulfur dioxide. Coal production has caused the subsidence of 1.30-million hectares of land, and less than 62% of the land can be cultivated again. About 34 billion m^3 of gas was emitted from mining in mines, among which 14 billion m^3 was extracted and 6 billion m^3 was utilized.[10] Coal combustion leads to a number of pollutants, such as oxynitride, sulfide, mercury, dust, black carbon and carbon dioxide.

---

8 National Bureau of Statistics of the People's Republic of China, *Discussions about Communiqué 2014: What Does 4.8% Decrease in Energy Consumption Per Unit of the GDP Mean?* http://www.stats.gov.cn/tjsj/sjjd/201503/t20150308_690781.html.

9 The Research Group of the China Coal Research Institute for "A Study of the Plan and Policy for the Total Quantity Control of Coal Consumption in China". A Study of the Development of the Coal Industry Based on the Total Quantity Control of Coal Consumption. 2015-01.

10 Wang Qingyi (ed.), *Energy Data 2013*. 2013.

According to relevant statistical information, among the main pollutants discharged in 2012, sulfur dioxide, oxynitride, soot, industrial wastewater, solid wastes and carbon dioxide were 21.18 million tons, 23.38 million tons, 12.26 million tons, 400 million m^3, 4.43 million tons and 8.8 billion tons, respectively. More than 60% of the above pollutants came from coal mining and its utilization. According to the China National Coal Association, 1,067 people died or were missing and the death rate was 0.29 people/1,000,000 t in coal mines in 2013. Funnels, the fall of the water level, water pollution and other problems caused by continuous massive coal mining caused great damage to water resources. Take Shanxi as an example: each mining of 1 t of coal destroys 2.48 t of underground water, and coal mining will pollute drinking water on the ground.

4.2 The Total Quantity Control of Coal Consumption Will Produce a Good Synergistic Effect on the Environment and on Health

Mercury from coal combustion accounted for more than 40% of mercury emissions nationwide. Heavy metals such as mercury have severely polluted the soil. In the current PM2.5 air pollution in various regions, the rate of primary PM2.5 contribution from coal utilization and combustion was 62%, while that of secondary PM2.5 contribution from coal utilization and combustion was 50%–61%, depending upon the region, being 56% on average.[11] Extensive air pollution is closely related to coal combustion and its utilization and the emission of industrial waste gases. Besides direct pollutant emissions, chemical pollutants from coal utilization are the important chemical precursors for the PM2.5 formation. In order to improve public health and reduce air pollution, it is necessary to eliminate pollution from coal smoke first, which is one of the main measures with a low cost and a quick effect. Controlling total coal consumption can bring about a very significant synergistic effect in protecting the environment and improving health. According to the report *Real Costs of Coal in 2012*,[12] the environmental and health cost incurred by coal development, transportation, combustion and utilization was 260 yuan/t, the loss and damage from carbon dioxide emissions was estimated to be about 160 yuan/t, the total cost was 457 yuan/t or higher. In 2012, 3.53 billion tons of coal were consumed, and the total real cost was 1.6132 trillion yuan/t, accounting for 3.2%

---

11 The Research Group of the China Coal Research Institute for "A Study of the Plan and Policy for the Total Quantity Control of Coal Consumption in China". Coal Contribution to Air Pollution. 2015-01.

12 The Research Group of the China Coal Research Institute for "A Study of the Plan and Policy for the Total Quantity Control of Coal Consumption in China". *Real Costs of Coal*. 2015-01.

of the total GDP in 2012. More recent studies will show that the environmental and health benefits from the total quantity control of coal consumption will become increasingly significant. In the current mechanism, environmental taxes and charges are only 30–50 yuan/t, most of which are concentrated at the end of production, while the pollutant charge at the consumption end is only 5 yuan/t. An environmental tax and a carbon tax have been put on the agenda.

4.3 Total Coal Consumption Should Be Subject to Restrictions from the Ecological Red Lines

In the past several decades, China's economy has rapidly developed at the expense of its resources and the environment. The important step for establishing a system of an ecological civilization is that ecological red lines should be designated so that future economic development will be subject to restrictions from them. The above action plan sets forth the goals for reducing the PM2.5 concentration in various regions by 2017 and stresses that these goals must be completed. According to the standard of the World Health Organization, three goals for the concentration of PM2.5 in the period of transition are an annual average of 35 µg/m^3, 25 µg/m^3 and 15µg/m^3 respectively, while the air quality criterion is an annual average of 10 µg/m^3. In our opinion, the national goal for reducing the PM2.5 concentration should be the recommended value of the first stage specified by the World Health Organization (35 µg/m^3) by 2025, and then the recommended value of the second stage (25 µg/m^3) in the next ten years, and that the air quality level will become equivalent to that of the developed countries (10–15 µg/m^3) in 2045–2050; this should be the red line for air quality. In addition, water resource challenges for China will become more severe. The per capita water resource quantity in China is 25% lower than the average world level. The water resources in North China are 1/10 of the average national level. China's coal resources and water resources feature a reverse distribution—coal resources in the central region and the regions close to West China account for 73.6% of those across China, while water resources in these regions make up 22.7% of those across China.[13] Some local regions subject to a shortage of water resources are the main coal bases, power generation bases and coal chemical bases in China, so water resources in these regions are excessively exploited. Therefore, the strictest water resource standard should

---

13   The Research Group of the China Coal Research Institute for "A Study of the Plan and Policy for the Total Quantity Control of Coal Consumption in China." Carry Out the Strictest Red Line Requirement for Water Resources, Restrict Coal Development and Utilization. 2015-01.

be resolutely implemented in North China and Northwest China, where the red lines for energy utilization, especially coal utilization, should be specified. Imposing restrictions on water resources can also help intensify water saving measures in coal development and utilization projects. According to a report from Tsinghua University and the Natural Resources Defense Council, the water saving effect from energy conservation in the 11th Five-Year Plan period was equivalent to the amount of water in the Yellow River.[14] Therefore, energy conservation should be closely combined with water saving, and both energy conservation and water saving are the basic constraining conditions for national economic development. A report from the China Institute of Water Resources and Hydropower Research specifies the rigid constraining requirements for water resources in the main coal-consuming industries and regions, and states that coal development should be based on water resources. According to this report, if the original pattern of coal development is continued, the existing water consumption and the strict water saving mode will not be able to satisfy the water resource supply. Only when the water consumption mode is in the coal control scenario and under the strict water saving mode can the water resource supply be satisfied. The requirement that coal development should be based on water resources must be strictly satisfied in 14 coal bases and North China, especially Northwest China.

4.4 Climate Change Requires Coal Consumption to First Reach Its Peak

China exercises the total quantity control of coal to reduce carbon dioxide emissions. China is the largest carbon dioxide emitter in the world. In 2013, China's carbon dioxide emissions amounted to about 8.7 billion tons and China's per capita carbon dioxide emissions exceeded the per capita level of the EU.

If the current emission trend goes unchanged, as forecast by the United Nations Environment Programme, by 2020, subject to 40%–45% reduction in China's carbon intensity, the scale of emissions will be close to the sum of the current emissions in all OECD countries—energy-related carbon dioxide emissions in the OECD were 12.34 billion tons in 2011.[15] The Chinese Government's strategies, policies and measures for addressing climate change are the focus of people's attention. The total quantity control of coal consumption is a

---

14 "Water Saving Effect from Energy Conservation in the 11th Five-Year Plan Period," a report from Tsinghua University and NRDC. 2015-01.
15 The Research Group of the China Coal Research Institute for "A Study of the Plan and Policy for the Total Quantity Control of Coal Consumption in China." An Analysis of Restrictions from Carbon Emission on Total Quantity Control of Coal Consumption and Synergistic Effect. 2015-01.

cutting-edge means for China to reduce its carbon dioxide emissions. Climate change has made China subject to severe impact and heavy losses. According to China's Climate Change Adaptation Strategy, climate change has severely affected the sustainable development of China's economy, its water resources, food security and public health. Specifically, grain yield in China may decrease by more than 25% by the end of the 21st century. Before reaching the peak of its carbon dioxide emissions, China should reach the peak of its coal consumption in advance. China should unveil its plan to the whole world in the negotiation of the new climate change agreement. If the total quantity of coal can be effectively controlled and reduced, the peak of China's carbon dioxide emissions will change. China's coal consumption decreased by 2.9% in 2014; as a result, carbon dioxide emissions in 2014 declined by 1% compared with 2013.

5 The Total Quantity Control of Coal Consumption Gives Rise to a More Technical Kind of Progress, and Frees the Sectors and Coal Bases from Dependence on Coal

5.1 *The Total Quantity Control of Coal Consumption Can Enhance the Efficiency of the Energy System*

Energy conservation is one of the priorities in China's energy strategy. Since the 1980s, the implementation of the energy conservation goals has ensured rapid economic development with less energy consumption in China. According to our estimation, if the structural pattern of the world's average energy consumption is adopted and other conditions remain unchanged, the efficiency of China's energy system can be improved by 18%–20%. As its coal efficiency is low, more funds and technologies are required in order to enhance its efficiency of coal-dominated energy. If cleaner energy is used, the efficiency of the end energy will be higher.

According to the report *China's Energy Flow and Coal Flow Diagram 2012*,[16] if high-quality energy sources, such as natural gas and renewable energy sources, are used in the energy system, the efficiency of the energy system can be improved in energy type, conversion and application. The efficiency of the energy system was 25.9%, 32.1% and 36.1% in 1980, 1995 and 2012, respectively. The efficiency of the coal system was 31.2% in 2012; obviously, the excessively high coal consumption led to the inefficiency of China's whole energy system.

---

16 The Research Group of the China Coal Research Institute for "A Study of the Plan and Policy for the Total Quantity Control of Coal Consumption in China." *China's Energy Flow and Coal Flow Diagram 2012*. 2014-12.

With the coal control plan, the efficiency of the energy system is estimated to reach 38.5%, 40.2% and 44.1% by 2020, 2030 and 2050, respectively. China can catch up with the developed countries of the late 1990s in terms of the efficiency of its energy system by 2030–2040 and keep pace with the developed countries by 2050.

In the total quantity control of coal consumption, energy conservation is considered an extra requirement. The total quantity control of coal consumption means a number of requirements for the main energy-consuming enterprises, so it is necessary to stimulate enterprises to better tap their technical potential and improve the efficiency of their energy conservation in order to achieve the goal for the total quantity control of coal consumption. In the 13th Five-Year Plan period, it will be indicated that other external forces will be essential for promoting energy conservation. Besides investments and technologies, the mandatory goal for the total quantity control of coal consumption can further push forward energy conservation. The emissions of carbon dioxide and other pollutants from the transformation of the energy structure can be reduced by 18%–19%. In 2014, the annual intensity of energy conservation reached all-time high, 4.8%, thanks to the reduction in China's coal consumption.

5.2 The Total Quantity Control of Coal Consumption Can Speed up the Removal of Excess Production Capacity in the Sectors with High Energy Consumption

Since 1949, total coal consumption has accounted for more than 65% of the total energy consumption. China has carried out a strategy for coal-based energy development and has thus heavily depended upon coal for a long time. In many industries with high energy consumption in China, such as the steel, cement, electrolytic aluminum, plate glass and ship industries, only about 75% of the production capacity is utilized, while about 25% of the production capacity is in excess, resulting in a huge waste of energy and investments. For example, in the steel industry, coal consumption makes up more than 70% of the total consumption. If we can reach the goal for the total quantity control of coal consumption in industries with high energy consumption, we can cut down, phase out, dispose of and remove the excess production capacity, and improve the energy efficiency and the competitiveness of the industries and sectors with high energy consumption at home and abroad. Moreover, these sectors can break away from their heavy dependence on coal in this process and concentrate their efforts on industrial structural adjustments, product quality and energy substitution. The adoption of new, efficient and clean energy technologies can further give an impetus to industrial development. It can be predicted that the peak of the industries with high energy consumption is closely related to that of coal consumption, and the total quantity control of coal consumption can further promote the early appearance of the peak in the industries with high energy consumption.

5.3 The Total Quantity Control of Coal Consumption Pushes Forward Reforms and the Sustainable Development of Coal Bases

Rapid economic development aroused a huge demand for energy; rational exploitation and utilization of coal resources were overlooked in the coal sector. Coal was massively exploited at the expense of the environment and resources in order to meet the huge demand for energy; as a result, the coal sector was overwhelmed by its debts and failed to repay them for several decades. China's main coal-producing provinces, such as Shanxi, Inner Mongolia, Shaanxi, Ningxia and Guizhou have relied upon coal production and supply in order to impel the local economic development. With the adoption of the developmental path for several decades, these energy bases have faced grave challenges in their economic transformation and their dependence on coal has severely restricted the local developmental lines of thought, so the previous economic adjustments and reform efforts have become fruitless. In the "golden" period of coal development of the previous ten years, coal bases lagged behind other regions economically; investments and technology R&D were diverted to the coal sector in a deformed way. Amidst the current period of sluggish coal consumption, the people in these coal bases deeply reflected on the past, recalled the painful experience and decisively made changes. Under the plan for the total quantity control of coal consumption, the authorities in coal-producing provinces have planned a scientific and sustainable production capacity, and set the goal for controlling the total coal production with the aim of shaking off their dependence on coal.

5.4 The Total Quantity Control of Coal Consumption Calls for Establishing an Ecological Compensation Mechanism

In the past several decades, 14 billion tons of coal were produced in Shanxi Province, of which 10 billion tons were externally transferred to support the coal demand and economic development across China. Coal production in Shanxi Province was mostly realized by the disruption of the local environmental ecology and a relatively high rate of deaths among the miners. In Shanxi Province, coal mining did not take into account how to increase the recovery rate, nor how to rationally protect and utilize resources. According to relevant green GDP calculations from the Chinese Academy for Environmental Planning, the green GDP in Shanxi Province over the years has been negative;

in other words, the environmental and resource losses caused by mining 14 billion tons of coal much exceeded the economic income in Shanxi Province. In such a circumstance, actions should be taken to control the total coal consumption, establish a rational and effective ecological compensation mechanism, ensure opportunities for the provinces where coal bases are located in order to recuperate and build up strength, and develop plans for a rational production capacity. The authorities in Shanxi Province recently vowed to control the coal production and shift investments to the non-coal industries, and to find new strong economic growth points. Only when the issue of coal demand is controlled can the issue of coal production be rationalized and established accordingly. We should protect the natural resources of the provinces where coal bases are located according to plans for a rational production capacity. In recent years, coal has been vigorously developed in Inner Mongolia, causing too much damage to the steppe ecology and the enormous water resources. The restoration of vegetation and an ecological environment entails a long-term heavy price. The establishment of a compensation mechanism and the setting of a recuperation time are helpful for deepening reforms and for fostering a rational and sustainable development in the coal sector. Coal at the consumption end is controlled, so the total quantity at the production end should be controlled, and disorderly and cutthroat competition at the expense of environmental disruption should be avoided in the production bases with excess production capacity.

6   New Explanation of the Perspective of Energy Supply Security

The total quantity control of coal consumption helps reexamine the perspective of energy supply security. In 2014, 308 million tons of crude oil were consumed in China, the degree of China's dependence on foreign crude oil was 59.5%. Fifty-nine billion m^3 of natural gas and 285 million tons of crude coal were imported to China and the degrees of China's dependence on foreign natural gas and crude coal were 32.2% and 8% in 2014 respectively.[17,18] The total degree of China's dependence on foreign energy was about 17%. As the supply and demand on the global oil and gas markets change, the increasing development and self-sufficiency of shale oil and gas in the USA are reshaping the world's energy landscape. China can fully utilize both global oil and gas

---

17   "Both Coal Imports and Exports Declined in 2014," *China Energy News*. 2015-01-26.
18   "Rising Dependence on Foreign Oil Tests Energy Security," *China Energy News*. 2015-02-02.

markets to change the structure of its domestic energy supply and consumption so as to guarantee environmental protection and a sustainable utilization of resources.

Furthermore, given the grim situation of air pollution at the present stage, China should increase its importation of gas, including pipeline gas and liquefied natural gas, from the international market. China has no fundamental conflicts of interest with the oil-supplying countries in the international community; on the contrary, China shares interests with them. However, the threats to energy supply security still exist. It is necessary to take precautionary measures and make preparations, especially in view of the impact from the violent price fluctuation on economic activities. At the present stage, the energy structure should be smoothly transformed.

7 How to Achieve the Goal of Total Quantity Control of Coal Consumption

7.1 *The Exercise of the Total Quantity Control of Coal Consumption Requires the Adoption of Market and Economic Means*

The goal of coal control is set and distributed mainly by administrative means, while only when more market-based means are adopted can measures be effectively implemented and the wasting of manpower and financial resources be avoided. The domestic coal price is higher than that of the coal imported from foreign countries, which is sending a warning to China's coal industry at the level of competitive strength. The coal sector is a competitive market. From the perspective of the current level of productivity and labor wages in China, the domestic coal price should be lower than that in the developed countries. In the merger of coal enterprises, environmental standards and health requirements as well as the rate of deaths of miners should be taken as the rigid indicators; rational economic means should be adopted to close down, halt, and merge the coal production facilities under an unsound operation and switch to other productions, so as to form several hundred large coal production enterprises and enhance the market competitiveness in the industry.

The imposition of the coal resource tax on an ad valorem basis is very important for protecting the environment and increasing the effective utilization of resources. In the coal-producing provinces, the environmental disruption is huge, so some ecological compensation taxes should be levied at the consumption end and the coal-producing provinces should be subsidized. Coal-related ecological compensation taxes and a carbon tax could reduce the demand for coal and internalize the external costs in coal production and consumption, embodying the costs in the price.

As there is a goal for the total quantity control of coal, a coal quantity trading market could also be developed. If the total quantity control of coal is exercised, the quality requirements for coal will be very important; the market will automatically adjust the demand for high-quality coal and mitigate environmental disruption.

7.2 Develop a Mandatory System for Phasing out the Backward Coal-Fired Technical Equipment, Boost R&D and Utilization of Efficient Clean Coal Technologies and CCUS Technologies

At present, many advanced clean coal technologies have become or been close to being mature at home and abroad; for example, efficient pulverized coal combustion, clean coal gasification, and integrated gasification combined cycle (IGCC) technology are fully able to ensure that the air pollution emission indicators are lower than the current national standards for the thermal power industry during coal combustion, while the latest domestic clean coal technologies can generally bear comparison with gas power generation in terms of pollutants, except for carbon dioxide. In addition, the development of coal-based polygeneration also offers a systematic solution for the modernization of coal utilization. China had been advocating efficient clean coal technologies for several decades, but the effects were hardly due to insufficient motive power. If the goal of control and pollutant emission requirements are put forward at the consumption end, the application of efficient technologies can be promoted and enterprises will be more motivated and willing to allocate more funds and manpower to research and development. Efficient, clean and low-carbon utilization of coal is encouraged while enterprise competiveness is enhanced. Coal will remain the main source of energy consumption in China for a very long time yet to come. Even if the proportion of coal consumption in total energy consumption decreases, lower than 25%, the total coal consumption will still be huge in China. China must develop the Carbon Capture, Utilization and Storage (CCUS) technology in coal consumption. The development and utilization of CCUS technology can provide solutions for energy utilization for many developing countries. The developing countries still need local low-cost coal resources for a certain period of time in order to develop their economies, so CCUS technology is indispensable for various countries in order to address climate change. The total quantity control of coal consumption is bound to further enhance China's international cooperation. China should more actively engage in international cooperation in the development of alternative energy, energy conservation, clean coal technologies, CCUS technology and research on relevant policies.

7.3 *Improve and Push Forward China's Statistical Accounting System for Energy and Carbon Emissions*

China was often questioned for the accuracy of statistical data on energy, especially coal. The local, provincial and municipal statistical data on energy were greatly different from the national statistical data on energy, mainly due to the great difference in statistical data on coal. Since 2000, the difference between the local statistical data on coal and those at the provincial level, between the local statistical data on coal and those at national level has been on the increase, making it difficult to deal with the difference. The difference in statistical data will lead to many errors in policy research and decision-making, and it will, to a certain extent, affect China's energy planning, developmental strategy and policy-making. In order to bring forward a feasible plan for the total quantity control of coal, China should intensify its efforts to study methods for reducing statistical errors, so as to implement the total quantity control of coal consumption at the local level and propel the central and local statistical departments to jointly work on finding solutions. With the adjustment of energy data in 2014, the coal consumption difference—850 million tons—was reduced to about 260 million tons. Further verification and adjustment will greatly improve the reliability of the data.

8   Conclusion

China is energetically transforming its pattern of economic development, and the transformation of the energy supply and consumption structure will enable China to take an efficient, sustainable path towards economic development with a low level of pollution and high output. At the present stage, the total quantity control of coal consumption is crucial for succeeding in the nationwide campaign against air pollution. In recent years, the people in domestic, industrial and academic fields have reached a consensus: China should, in the 13th Five-Year Plan (2016–2020) and long-term planning, develop and carry out an efficient, market-based and mandatory goal and policy for the total quantity control of coal consumption across China, carry out an objective responsibility management and implement a roadmap to accelerate the reduction of coal consumption and make coal consumption clean in a more rapid way.

CHAPTER 4

Revisiting the Private Automobile Ownership Debate after 20 Years

Shi Jian*

Abstract

Twenty years ago, the scholars Zheng Yefu and Fan Gang engaged in a famous debate on whether private cars should be vigorously developed, which involved a series of major issues including the choice of a pattern and lifestyle of economic development and the bearing capacity of energy environment; that debate had an extensive repercussion and much participation. Nobody could identify the winner or the loser, the superior side and the inferior side in the debate; however, the consequences on the economy, the environment, transportation and on society caused by the impressive 20-year development of the automobile industry are evident. What is the road ahead? This is a tough question for today's decision-makers and consumers.

Keywords

Zheng Yefu and Fan Gang's debate – social reality – policy adjustment

The famous debate on cars occurred in the 1990s. Nobody has assessed who the winner was, but most of the problems involved in that debate have become reality after 20 years, so people have to think about them again.

1   Initiation of the Debate

An article entitled *Criticism of a Car Civilization* written by Zheng Yefu, professor of the Department of Sociology, Renmin University of China and published

* Shi Jian, Editor of Chinadialogue's Beijing Office, participated in writing a number of articles on the environmental and energy issues.

in the *Guangming Daily* on August 9, 1994 sounded the bugle of that debate. In that article, Zheng Yefu objected to the ownership of private cars by Chinese households, and believed that the wise choices in Chinese cities in the future would be buses and bicycles rather than cars. However, that article quickly came under attack. An article entitled *Criticism of "Civilization Criticism"* was published in the *Guangming Daily* on November 8, 1994. The article was written by the economist Fan Gang.

In his article, Fan Gang took the perspective of an economist to discuss the car civilization on six aspects and analyzed Zheng Yefu's article point by point. The six aspects involved the charm of civilization, the extreme limit of development, the paradox of human nature and the sorry-ness of backwardness. Their views contrasted sharply with each other. The important view of Fan Gang was that the car system was an important support for the growth of employment, income and the national economy, and he denied that the automobile industry would hinder development.

As expected, the attack from Fan Gang quickly received response from Zheng Yefu. Twenty-four hours later, Zheng Yefu wrote another long article to refute Fan Gang's view. He responded that the first advantage of a backward country was that it should not necessarily repeat the mistakes made by the advanced countries, but it should learn from them. Therefore, he doubted whether China could take an industry which was being questioned and would engulf most of China's energy sources as its economic cornerstone.

The editor of *Guangming Daily* commented, "as shown by both articles, the choice of the method of car development is not only the choice of a consumption mode and lifestyle, but also the choice of a way of thinking. It involves the following underlying questions: What is civilization? How should we consider progress? How should we look at development, even the future mode of human development and its ultimate goal as well?" The central issue of whether cars should be accessible to Chinese households also involved opinions from many domestic scholars. In 1995, the book entitled *A Debate on Cars* edited by Zheng Yefu came out, in which relevant articles on this subject were collected.

The famous economist Mao Yushi said that, when we discussed whether private cars should be developed, in fact, nobody raised any absolute objections, and nobody agreed that we should recklessly develop private cars; the substance of that debate consisted of the extent to which private cars should be developed, or more precisely, whether the extent of it should be determined by the market or whether the government should provide support or impose restrictions. At that time, China was certainly a kingdom of bicycles, and the bicycle was also the main means of transportation around the world. Owning a car was unimaginable for most of the Chinese people. According to the *China*

Statistical Yearbook, in 1994, 192 bicycles were owned by every 100 Chinese households, while there were 2,054,200 private cars in total.

Both sides presented not a few arguments in that debate. Some excerpts are given here:

> Zheng Yefu: Carriages travelled 6 km per hour in New York in 1907, cars are still moving at this speed in today's New York. People questioned: Why did we invent the internal combustion engine? Why have we done this for so many years?
> Fan Gang: The hourly speed of carriages was 6 km in New York in 1907, now cars are running at this speed in New York; even if this is true, the first cause is that now the population is larger but the number of cars is not large in New York.
> Zheng Yefu: Do people chase after cars in order to make money, satisfy vanity or pursue a better social life?
> Fan Gang: In general, cars or "private cars" indeed represent a sort of civilization and bring special enjoyment to people. You cannot prevent people from seeking civilization.
> Zheng Yefu: Can we bear the fuel burden involved in the American car civilization? Even if China is able to purchase more oil in the 21st century, can the international market, even our entire earth, support the rise of a large car country with a population of 1.1 billion?
> Fan Gang: What the human being owns or potentially owns is not limited to the resources, boundaries, technical capacity and lifestyle currently known to us.
> Zheng Yefu: The middle of every street in Chinese cities is available to cars, while both sides are available to bicycles. Most of the streets/roads for cars are wider than or as wide as those for bicycles. Who pays for building and maintaining roads/streets? In the final analysis, that payment is made by taxpayers, and taxpayers mostly come from car and bicycle owners.
> Fan Gang: There is an objection against developing private cars, but the wealthy cannot be restricted from riding in cars owned by them; otherwise, the difference between those with cars and those without cars will be increasingly large, most of the people will always see that increasingly luxurious cars are traveling on highways while they are riding a bicycle.
> Zheng Yefu: Is the development of the automobile industry the ultimate goal? The car is a means of transportation and a tool used by human beings. If the excessive development of the automobile industry causes choking traffic, is this advisable?

Fan Gang: If you cannot find any alternative industry, people will always doubt that you hope that we will remain among the backward countries with bicycles and donkey carts forever and see the wealthy people continue driving their cars and rich countries continue using their advantages in order to benefit from us.

They expressed their views and expounded their reasons. The fact that more and more private cars are owned by people proves a sustained increase in the level of urban construction and people's living standard. However, as the number of private cars is increasing, some social problems occur, drawing the public's attention. Many problems mentioned by both sides have become reality in the 21st century, and the heavy PM2.5 pollution from fossil fuel combustion and automobile exhaust emissions is precisely a big concern for the Chinese people, not to mention such problems as road congestion and high taxes and dues which have existed for many years.

2   Who the Winner Was

In 2005, Beijing TV invited both scholars to reignite that debate which was launched ten years before. Zheng Yefu accepted the invitation with pleasure, but Fan Gang did not want to join in. Zheng Yefu considered Fan Gang's refusal as admitting to being afraid to accept the challenge. Zheng Yefu believed he was the winner. Let us look away from that debate and take a look at the reality of these past 20 years. Chinese car makers have increased their production and have paid no attention to that debate.

According to the latest *China Statistical Yearbook*, 105,016,800 private cars were available in China in 2013, more than 50 times those in 1994; based on a population of 1,360,720,000, as shown by the statistical data of 2013 (including both urban and rural areas), 8 out of every 100 people owned private cars. If single cities are considered, the situation is more surprising. Take Beijing as an example, according to the *Beijing Statistical Yearbook*, 3.11 million private cars were owned in Beijing in 2013, 14 out of every 100 people owned private cars.

As mentioned by Zheng Yefu in that debate, some people purchase private cars in order to show off; however, in the objective reality, some people purchase private cars in order to meet their needs. The huge functions of the public transportation system, including buses, tramcars, minibuses, subways and taxis, cannot completely replace some conveniences of private cars; for example, the transfer to another car or subway is inconvenient, and travelling with the elderly and children is unsafe.

The rapid development of private cars in China offsets, to some extent, Zheng Yefu's sense of achievement from that debate, and it even saddens him. Are his arguments incorrect or is social the development following a crooked road? Zheng Yefu also remembers a consultation meeting by experts in 1998 in his blog, in which the newly-elected deputy mayor Wang Guangtao made the following opening remarks: "I am available in this meeting throughout the day to listen to your opinions. However, you should not discuss the following two points: First, inhibit private cars; second, develop subways because they go beyond my authority." After he walked out of the meeting venue, many discussions occurred; many experts said that we cannot discuss the lasting development and stability of Beijing's transportation if we are not allowed to discuss the suppression of private cars and the development of subways.

Zheng Yefu also said that Zhu Rongji explained why the article entitled *Vigorously Develop the Public Transportation* published by him on February 1, 2003 was chosen as the last article in the book entitled *Zhu Rongji on the Record* at Tsinghua University on April 22, 2011. Zhu Rongji discussed China's national conditions, and believed that the urban population was huge, the bearing capacity of transportation was limited, the degree of dependence on the importation of foreign fuel was high and exhaust pollution was severe, so he did not agree that everyone should buy a private car; on the contrary, he thought that cars should be properly developed and the government should not spend much money to subsidize and promote them. China should energetically develop public transportation, such as buses, urban light rail transit, even high-speed maglev trains. "In general, we must concentrate more efforts and attention on developing public transportation; we should not focus on developing cars!" Zheng Yefu did not stop criticizing the development of private cars and advocating other means such as public transportation after Fan Gang refused to further engage in that debate.

In 1995, he wrote an article and proposed to replace the official cars for cadres below the bureau level with a transportation subsidy; he stressed that citizens would never enjoy a high-quality public transportation system if officials were not included. In 1996, he suggested building special lanes for public transportation. In 2004, he reversed his previous opinion on inhibiting private cars, and pointed out that the government took sides with those with private cars in terms of road construction costs and thus encouraged a car-buying spree. In 2005, he wrote an article to give more details about this issue, and suggested that those people with private cars bore only 19% of Beijing's street construction costs, which was a serious policy bias.

In order to solve the traffic jam problem, it is imperative to build more streets/roads and flyovers which are highly costly projects. If project costs are not borne by those driving private cars, they will be borne by other ordinary people. This is unfair and encourages people to buy cars, causing exceptional street/road congestions because it is not necessary to pay for the use of streets/roads.

Before the fuel oil tax was introduced in 2008, he said that it was essential to levy heavy taxes on the consumption of scarce materials, and it was completely impossible for 1 yuan/L fuel oil tax to signify a resource shortage to society. It was lower than that in EU countries (about 6 yuan/L) and our surrounding countries and territories, only close to that in the USA. However, we cannot imitate the American lifestyle because we do not have so many resources. Twenty years later, the significance and value of that debate still exist. Now many problems mentioned by Zheng Yefu have really occurred. Social development cannot avoid these problems, and Zheng Yefu says that he still holds his views and firmly believes in them.

Twenty years later, this debate is no longer heated, but the issue concerning private car development has drawn the attention of more people. Nobody can judge who the winner was in that debate and history will not come to a halt, but the way of thinking and many doubts provoked by it have left true trails in history.

3 What the Road Ahead Is

Many adjustments have been made in the whole of society to solve a series of problems in car development though they appear to be passively made. In recent years, given the increasingly severe air pollution and more and more traffic jams, such measures as regulation and control of car increment and imposition of restrictions on using cars based on odd-even numbers on license plates have been adopted, even a stricter order of purchase restriction has been issued, in many cities including Beijing, Tianjin and Shanghai.

During the APEC Meeting in 2014, strict measures for restricting the use of cars were taken in many areas in North China, where citizens were encouraged to ride on public means of transportation. These measures produced marked effects, streets/roads were not jammed and haze was temporarily removed. However, in order to fundamentally address the issue of exhaust fume pollution from cars in large quantities, we should focus on developing public transportation. It is absolutely necessary and highly feasible to vigorously promote clean public transportation in cities; in order to ensure that citizens can enjoy convenient public transportation, giving priority to it has become the inevitable choice for urban development.

CHAPTER 5

Soil Remediation: Still a Long Road Ahead

Liu Hongqiao*

Abstract

With the release of the *Communique of the National Survey on the Status of Soil Pollution*, soil remediation became a hot topic in 2014. However, severe pollution and a clear-cut stand taken by the central leadership did not turn the soil remediation market into a booming one. In fact, there are disagreements on the thinking regarding legislation; laws, regulations and policies are not explicit; no industrial and technical standards are available; financing modes are not diverse; it is difficult to guarantee funds for remediation; the technical route is unclear; as a result, the capital market becomes hesitant towards soil remediation. Meanwhile, the Changsha-Zhuzhou-Xiangtan pilot project in Hunan Province for the remediation and comprehensive treatment of the cultivated land polluted by heavy metals, as funded and initiated by the central finance, continues being carried out under the state-led "exploration" mode; its treatment plan is controversial. The romantic imagination about the prospect of soil remediation is still subject to the stark reality: there is an entanglement of multi-party interests.

Keywords

status quo of soil pollution – status quo of soil remediation – controversies in the *Soil Pollution Prevention and Control Law* – soil remediation market

On April 17, 2014, the Ministry of Environmental Protection and the Ministry of Land and Resources jointly released the *Communiqué of the National Survey on the Status of Soil Pollution* (the "Communiqué"), lifting the real veil from China's soil pollution for the first time.

According to the Communiqué, the status of China's soil was not optimistic—the overall national rate of soil exceeding the standard was 16.1%;

---

*   Liu Hongqiao, Research Fellow at "China's Water Crisis" and "chinadialogue" cooperative water projects, worked at *Nanfang Metropolis Daily* and Caixin Media; his studies focus on pan-environmental, health, scientific and technological issues.

the rate of the point locations of the cultivated land exceeding the standard was 19.4%; the rate of forestland, grassland and unused land exceeding the standard was higher than 10%.

The methods adopted and the results obtained from this survey are controversial in the academic community and in the industry, but the Communiqué provides an overview of unprecedented soil pollution in China. Both ministries frankly stated, "The soil in some regions is highly polluted, the environmental quality of the cultivated land and soil is worrying, there are obvious problems in the soil environment of the industrial and mineral wasteland."

The Communiqué highlights the grim situation of soil pollution in China and also generates the romantic imagination about soil treatment—the Ministry of Environmental Protection will introduce the *Action Plan for Soil Pollution Prevention and Control*, a soil remediation market with a budget of 100 billion yuan in the short term and 1 trillion yuan in the long term could be expected, and soil remediation will become the most vibrant force in China's environmental protection industry.

These guesses are not groundless rumors. In "two sessions" (of the NPC and the CPPCC) in early 2014, the State Council vowed, in the government work report, to resolutely declare "war" on pollution. In March, Zhou Shengxian, former Minister of Environmental Protection, decomposed this "war" and pledged to fight three hard-fought "battles"—prevention and control of air pollution, water pollution and soil pollution. However, severe pollution and a clear-cut stand taken by the central leadership did not make the soil remediation market brisk. As a matter of fact, laws, regulations and policies are not well-defined; there are no industrial and technical standards; financing modes are not diverse; it is difficult to guarantee funds for remediation; the technical route is unclear, so the capital market shows hesitation towards soil remediation. Meanwhile, the Changsha-Zhuzhou-Xiangtan pilot project in Hunan Province for remediation and comprehensive treatment of the cultivated land polluted by heavy metals, as funded with 1.15 billion yuan by the central finance and initiated in late April 2014, continues being carried out under the state-led "exploration" mode; its treatment plan is controversial.

1   The Status Quo of Pollution

1.1   *The* Communiqué of the National Survey on the Status of Soil Pollution *Preliminarily Uncovers the Real Situation of Pollution*

In July 2006, the Ministry of Environmental Protection and the Ministry of Land and Resources launched the first survey on soil pollution nationwide.

During this survey, soil pollution and the resulting impact on the environment and health gradually drew public attention; however, when the general public applied for the release of the data from the general survey on soil pollution, it was informed that these data were a "state secret."

Eight years later, in 2014, the two ministries finally released the *Communiqué of the National Survey on the Status of Soil Pollution*, making public this "state secret." According to the Communiqué, agricultural, industrial, forest and unused land was polluted to varying degrees. The rate of the point locations of cultivated land that exceeded the standard was 19.4%, higher than the national average rate, which was 3.3%; the point locations subject to slight, mild, moderate and heavy pollution accounted for 13.7%, 2.8%, 1.8% and 1.1% of the point locations exceeding the standard, respectively.

In this survey, the two ministries also conducted research focused on the pollution in typical land parcels and the surrounding soil. As indicated, with respect to typical land parcels including the land used by heavily polluting enterprises, industrial wasteland, industrial parks, the land for centralized disposal of solid wastes, oil-producing areas, mining areas, sewage irrigation areas and arterial highways, the point locations exceeding the standard on these typical land parcels accounted for as high as 20.3%–36.3%.

However, this soil survey is a general one with sparse sampling point locations, thus it only provides an overview of the environmental quality of China's soil at the macro level. Take cultivated land as an example: one point location was arranged in a grid of 8 km x 8 km—6,400 hectares or 96,000 mu (1 mu = 0.0667 hectares) sectors, which was equivalent to only two point locations in all of Beijing. In spite of this, the survey result was shocking.

1.2   Pollution Increased Continuously in Various Types of Soil

The saying that "new debts occur when old debts have not yet been repaid" is appropriate for describing the current situation of soil pollution in China. At the present stage, the total amount of pollution emission from such sectors as industry, mining, agriculture and forestry in China is still massive, and the new pollution in various types of soil is still on the increase, and the situation of soil pollution prevention and control is grim.

Take heavy metal cadmium as an example: Compared with the 7th Five-Year Plan period, the content of cadmium in the soil generally increased nationwide; the increase in Southwest China and in the coastal areas exceeded 50%, while that in North China, Northeast China and West China was 10%–40%.

Among the various types of land utilization, the land used by heavily polluting enterprises and the surrounding soil saw the worst case of exceeding the standard. In this survey, with regard to the land destined for heavily polluting enterprises—including those specializing in ferrous metals, nonferrous metals, leatherware, papermaking, oil and coal, chemical and pharmaceutical engineering, chemical fiber and rubber, mineral products, metal products, electric power—and the surrounding soil, the point locations exceeding the standard accounted for 36.3%.

Besides industrial pollution and mining pollution, improper agricultural production activities carried out by human beings continued to pollute the cultivated land. Both ministries recognized that the irrigation with waste water, irrational use of agricultural inputs including chemical fertilizers, farm chemicals and agricultural films, and livestock breeding were the main causes for soil pollution of the cultivated land.

1.3 The Impact on the Environment and on Human Health Has Become the Focus of Public Attention

While releasing the *Communiqué of the National Survey on the Status of Soil Pollution*, the two ministries listed three great hazards from soil pollution.

The first hazard was the impact on the output and quality of agricultural products. It included crop failure, quality damage and loss of economic benefit. Moreover, long-term eating of polluted agricultural products may severely jeopardize human health.

The second hazard was the impact on the safety of human settlements. Soil pollution of the land destined for residential and commercial construction may endanger the health of human beings in various ways, including oral and nasal inhalation and skin contact. The pollution sites directly used for development and construction without treatment may cause long-term damage to people.

The third hazard was the threat to the safety of the ecological environment. Soil pollution can damage the normal functions of soil and may also be converted and transferred to enter surface water, underground water and the atmospheric environment, affecting other environmental media and threatening the safety of the sources of drinking water.

2 The Status Quo of Remediation

2.1 Exploration of Large-Scale Soil Remediation and Treatment in the Changsha-Zhuzhou-Xiangtan Region

A real challenge for China is that there are no economically feasible measures in the world with respect to the extensive areas of polluted cultivated land. Japan made huge investments in the course of 33 years; it finally took the improved soil from other places to control the heavy metal pollution in the Jinzū River basin, but many problems could not be removed. In the USA, with regard to hundreds of thousands of land parcels under the Superfund, only a very small proportion of heavily polluted sites were remediated, while enormous areas of polluted cultivated land were left unused, or remediated by introducing plants and organisms at relatively low costs which involves a long remediation cycle.

Regarding China, the amount of polluted cultivated land is too extensive to take the improved soil from other places; China's population is huge and the land is averaged to each household, so it is difficult to carry out plans for the majority of the abandoned polluted land. China must independently explore its own method.

In April 2014, the Ministry of Finance and the Ministry of Agriculture declared the initiation of the remediation and comprehensive treatment of the cultivated land polluted by heavy metals all across China, and they first launched a pilot work project in the Changsha-Zhuzhou-Xiangtan region of Hunan Province. This pilot project covers 1.70 million mu and is planned to be carried out within 3–5 years. Only in 2014, 1.15 billion yuan was allocated from the central finance, and an investment was also made accordingly by Hunan's finances.

So far, this is the polluted cultivated land treatment project that has the largest area of treatment and relatively complicated pollution on the largest investment scale. Previously, the governments at various levels led pilot remediation of the cultivated land in Shiping Village, Gulin County, Sichuan Province, pilot treatment of soil pollution within the sweet cherry base in Zongqu Village, and the cultivated land treatment project in Wanshan District, Tongren City, Guizhou Province, and some tasks of farmland soil or river basin pollution remediation and treatment in Dongdagou in Gansu Province, Wuxing Village in Jiangxi Province, Heqiao Village in Jiangsu Province, Linxiang City in Hunan Province, Fenghuang County in Hunan Province and Dahuan River in Guangxi.

According to relevant plans, the Ministry of Agriculture will perform classified management of the pilot polluted farmland in the Changsha-Zhuzhou-Xiangtan region. Based on the degree of soil pollution and cadmium content in rice, 1.70-million-mu of pilot cultivated land will be classified into slightly polluted areas, moderately polluted areas and heavily polluted areas. In the slightly polluted areas, agricultural measures such as low-cadmium varieties, rational irrigation and acidity regulation will be adopted jointly. In the moderately polluted area, special varieties will be used, a plantation system will be introduced in the special zones, purchases will be made by special enterprises, storage will be performed by special factories and agricultural products will be under closed operations. In the heavily polluted area, substitute plantations of crops will be carried out; the crops which are not directly eaten and are not food, such as cotton, silk, mulberry, hemp and flowers, are planted.

However, this pilot project under the Ministry of Agriculture does not directly involve the remediation of the polluted cultivated land, but it is designed to control the channels of pollution that affect crops and ensure that the polluted crops are not eaten by people.

According to soil remediation experts, in the short term, such comprehensive control measures can control the environmental and health impact from the polluted cultivated land at relatively low costs, but in the medium and long term, this method for remediating and treating the polluted soil actually treats the symptoms instead of removing the root causes—the polluted soil is a source of pollution and can pollute the ecosystem, including underground water and surface water.

In fact, before the *Action Plan for Soil Pollution Prevention and Control* was introduced, it was difficult for government departments to reach a consensus on the goal for remediation and treatment of the polluted cultivated land. Should the pollutants be removed from the soil to "purify" the cultivated land or should relatively mild methods be adopted to control pollution within an acceptable scope so as to ensure the safety of agricultural products? The competent departments at various levels hold greatly differing views.

2.2 Problems Relating to the Remediation of Polluted Sites and Mining Areas Have Loomed Large

Unlike the remediation and treatment of the polluted farmland, the remediation of polluted sites and mining areas is characterized by a relatively high degree of marketization, large-scale investment and diverse sources of funds. Based on open channels, the database of the Xiangsu (Yixing) Institute of Environmental Industry has offered statistical data on the soil remediation projects since 2006. Among 358 soil remediation projects completed through the actions of the market, the remediation projects of the polluted sites accounted for 50%, while the average amount of their investment greatly exceeded that of the remediation projects regarding mining areas and cultivated land, and reached 53.35 million yuan.

The remediation and treatment projects of polluted urban sites which are underway, mostly involve the historical land left by state-owned enterprises in the wave of suppressing secondary industry and developing tertiary industry. Some polluters of the polluted sites are unable to pay the remediation costs and some of them cannot be found. However, thanks to the increasing appreciation of the land in developed cities, it is possible to conduct the costly remediation of the polluted urban sites. The ultimate beneficiaries—real estate developers and local governments—are willing to directly or indirectly pay for the remediation of the polluted urban areas. However, with the stagnation of the real estate market and its increasing saturation in the main developed cities in 2014, these driving forces for investments are on the decline.

The governments have actively participated in the remediation projects of the polluted sites. According to the 12th Five-Year Plan, 30 billion yuan will be allocated from the central finance to remediate the polluted soil all across China; 30%–45% of the financial aid will be provided by the central finance for the polluted urban land left over throughout history.

However, the involvement of government departments has not changed the situation; the top-level treatment design is imperfect, the overall national demand for treatment and the thoughts regarding treatment are unclear, and this has caused a number of problems.

At the macro level, the Central Government has not yet identified the overall national demand involving heavy metal pollution, so it is difficult to develop an overall line of thought about treatment; some special remediation plans or regional plans, and implementation plans do not serve any particular purpose and are not of any significance as a guideline.

When it comes to the selection, examination and approval of the projects, local governments have acted blindly and have not given these aspects any systematic and long-term consideration; they have even passively applied for projects when special fund subsidies from the central finance were available.

At the operational level, the remediation of the polluted sites and mining areas has often required an extensive preliminary survey and a careful evaluation of the risks involved, and the preliminary work took a long time. Local governments did not attach importance to the preliminary survey in most cases; delay occurred in allocating the funds from the central finance to the lower levels or the project construction could not be started after the funds were allocated, so the funds from the central finance were not used efficiently.

Since 2012, Shen Yiyang, senior energy and carbon financing expert from the Asian Regional and Sustainable Development Department of the Asian Development Bank, has participated in reviewing heavy metal pollution treatment projects many times. According to his observations, the investments from the central finance were made on the basis of a fixed proportion, so the scale of construction and the amount of investments were falsely increased in some projects for which an application was filed, and the treatment was even excessive in some projects and insufficient in others. Moreover, the overall efficiency of investment funds from the central finance was low, there was a large gap in local supporting funds, and a market-oriented mechanism of diversified investment and financing has not yet been established.

2.3 The Market was Opened at a High Level but Declined to Lower Levels, a Tens-of-Trillion-Yuan Market Has to Be Tapped

As early as December 8, 2013 when the China Listed Environmental Company Summit was held, Zhuang Guotai, Director of the Department of Ecology, the Ministry of Environmental Protection, said that, once the soil treatment market was opened, its scale might reach tens of trillion yuan. In March 2014, *China Environment News* instructed by the Ministry of Environmental Protection once again published an article stating that about 50 million mu[1] out of 150 million mu[2] of polluted cultivated land in China was moderately and heavily polluted cultivated land, and that merely remediating this cultivated land would require more than 8 trillion yuan.

However, the reality is grim. More than ten years have passed since April 28, 2004 when the construction workers in Songjiazhuang were poisoned—a hallmark incident of soil remediation in China. With more than ten years of development, the scale of the market for soil remediation in China is still far from the optimistic expectations of the Ministry of Environmental Protection.

According to China Environmental Remediation Online (www.hjxf.net), in 2013, the scale of the market for China's environmental remediation industry was only 6.7 billion yuan, the annual output value was 2 billion yuan, and the environmental remediation industry accounted for only 0.19% of the total scale of the environmental protection industry.

According to incomplete statistics, as of 2014, there were more than 1,000 registered soil remediation companies in China. However, as shown by a further classification of soil remediation enterprises, nearly 70% of the relevant enterprises were specialized in soil remediation engineering consulting and design; only 35 enterprises were principally engaged in equipment manufacturing though they have grown rapidly in the last two years in China and they mainly focused on the manufacturing of small equipment; most of the technology R&D enterprises in the industry were large environmental research institutions, while very few enterprises were capable of R&D.

---

1 It was an official estimate in the 1990s.
2 Ibid.

The government is the largest advocate in the soil remediation market at the present stage. As from early in the 12th Five-Year Plan period, China has attached importance to soil pollution prevention and control; it has introduced a number of policies and has invested special funds. In November 2009, the Ministry of Environmental Protection issued the *Guiding Opinions on Strengthening the Prevention and Control of Heavy Metal Pollution*. In 2011, the State Council officially approved the *12th Five-Year Plan for the Comprehensive Prevention and Control of Heavy Metal Pollution* and the *Plan for the Comprehensive Control of Heavy Metal Pollution in the Xiangjiang River Basin*. These three documents heralded the start of polluted soil remediation at the central level.

However, as the problems from government-led remediation and treatment projects have gradually surfaced, prudence has been exercised in terms of the fund input from the central finance. Meanwhile, since financing channels are not numerous, business models are not perfect, technical specifications are not available, and control responsibility and the means of payment are unclear, the investment of market capital in soil remediation has been interrupted by many worries.

In the opinion of Yu Xiaodong, Deputy Director of the Resources & Environmental Business Department, China International Engineering Consulting Corporation, the soil remediation industry at the present stage is similar to some basic construction fields of the 1990s-initially, there funds arrived from the central finance, and the market players believed that the demand was brisk, but it was difficult to understand the situation, finally the funds from the central finance exceeded the total market investment; as a result, the funds from the central finance were no longer massively allocated to be used in the area of soil, causing damage to the industry.

According to industry analysts, China's soil remediation industry is at the initial stage. Compared with the development path of the US soil remediation industry, China's current development level is approximately equivalent to the level of the USA during the mid-1990s. In the next ten years, as China's urbanization and industrialization reach a higher level, the environmental treatment market will really be opened.

"The overall situation of soil remediation is very good; it can be considered the largest virgin land in the future environmental protection industry. If it is carried out well in the future, many listed companies will emerge; otherwise, many enterprises will ruin their reputation," said Chen Tongbin, the Director of the Environmental Remediation Center of the Institute of Geographic Sciences and Natural Resources Research, Chinese Academy of Sciences.

3 No Breakthroughs in Top-level Design

The State Council has stated in the *Recent Work Arrangements for the Protection and Comprehensive Treatment of the Soil Environment* that, a policy, law, regulation and standard system for the protection of the soil environment will be established by 2015. However, major breakthroughs have not yet been made in soil legislation.

3.1 *Controversies in the* Soil Pollution Prevention and Control Law

The initiation of soil-related legislative work can be traced back to the Standing Committee of the 12th National People's Congress which, for the first time, incorporated the protection of soil environment into the first category of items under legislation planning. The legislative work was led by the Ministry of Environmental Protection which worked with the National Development and Reform Commission, the Ministry of Land and Resources, the Ministry of Agriculture and other departments to set up the leading group, working group and the corresponding expert group for drafting the law regarding the protection of the soil environment.

According to Ding Min, Deputy Director of the Proposed Law Office, the Environmental Protection and Resources Conservation Committee of the National People's Congress, as shown by the current schedule of legislation, the revision and improvement of the draft of the *Soil Pollution Prevention and Control Law* will be finished by the end of 2016, and it will be deliberated by 2017.

At present, various departments have numerous different views about the *Soil Pollution Prevention and Control Law*. There are two totally different views merely about the legislative purpose to be specified in the general provisions: Should the quality of the soil environment or the quality of agricultural products in the soil be guaranteed?

If the legislative purpose is to guarantee a healthy soil environment, that means a more extensive soil remediation or a more "thorough" engineering of pollution removal. If it is only necessary to ensure that agricultural products do not exceed the standard while the soil is allowed to exceed the standard, this means that the pollution control subproject in the Changsha-Zhuzhou-Xiangtan cultivated land remediation and treatment project may become the mainstream in the future treatment of polluted farmland.

Under the current government framework, the work of many ministries and commissions, including the Ministry of Environmental Protection, the Ministry of Land and Resources and the Ministry of Agriculture, involves the soil, so legislators need to coordinate the interests of ministries and commissions in assigning the responsibilities. Controversies still exist in assigning the responsibilities for specific work, such as the survey system, the standard, monitoring, evaluation and remediation funds and protection of the cultivated land.

It is often said in the industry and the academic community that there is no separation of water and soil, which means that both the soil and the underground water should be considered as an integral whole, and should not be separated, regardless of pollution or remediation. This leads to another controversy: Should the *Soil Pollution Prevention and Control Law* set forth the provisions for the pollution of underground water as well?

Another debate focuses on prevention and control of the soil pollution of the cultivated land. Because the prevention and control of the soil pollution of the cultivated land directly affects crop production, food safety and public health and no owner can be found in the case of lodging complaints about the pollution of much of the cultivated land, the financing mechanism, legal liability and remediation standard for the protection and remediation of the soil of the cultivated land become the main connections among the various interests.

3.2 *Policies and Standards are to Be Improved*

In parallel with the legislative work on the soil, the *Action Plan for Soil Pollution Prevention and Control* was developed and deliberated under the State Council. As early as March 18, 2014, this action plan was deliberated and adopted in principle at the executive meeting of the Ministry of Environmental Protection, and was submitted to the State Council. At that time, the Ministry of Environmental Protection suggested that this action plan should be released within the year 2014.

According to the Ministry of Environmental Protection, this action plan states that we will push forward the protection of the soil environment according to laws, resolutely cut off the soil pollution sources, carry out a classified management of the agricultural land and a classified control of the land destined for construction and the soil remediation projects. Its purpose is to ensure that the soil environment of agricultural land can be effectively protected and the worsening of soil pollution can be contained in China by 2020.

Before then, the State Council released the *Action Plan for Air Pollution Prevention and Control* to address air pollution, and the national administrative plan played a significant role in promoting soil remediation. With the previous *Action Plan for Air Pollution Prevention and Control*, the *Action Plan for Soil Pollution Prevention and Control* is expected to become the programmatic document for the remediation and treatment of polluted soil in the next stage, which will inject vigor into the soil remediation market.

Besides the *Action Plan for Soil Pollution Prevention and Control*, it is extremely urgent to improve the standard of quality regarding the soil environment and introduce a standard and a technical standard for soil remediation.

Moreover, there is no clear institutional framework for the developmental mode of the soil remediation industry. Various ministries and commissions adopt different developmental modes for the environmental protection industry—the Ministry of Finance favors the public-private partnership mode under which the government cooperates with enterprises, and risks and interests are shared; the National Development and Reform Commission adopts and emphasizes a marketization mechanism and the third-party governance mode under which funds are advanced and financing is conducted by a third party, payment is based on performance; the Ministry of Environmental Protection advocates integrated environmental services. It is still necessary to explore the industrial developmental mode which is most suitable to the future reality in China.

CHAPTER 6

Progress in China's Environmental Legislation in 2014

*Qie Jianrong**

Abstract

The amendment to the *Environmental Protection Law* was adopted in April 2004. It provides that punishment is imposed on a daily basis, seizure and attachment are introduced, and administrative detention is enforced against law breakers; it reinforces information disclosure and public participation, and specifies the qualifications of the subject for environmental public interest litigations, thus it is considered as the strictest environmental protection law. However, the prospects for the enforcement of this law have not been optimistic. In the actual legal environment, the problems that laws are not strictly enforced and lawbreakers are not punished are severe and lingering. Those prominent problems that the environmental protection law enforcement team has not made much progress, is unable to conduct investigations, takes no necessary actions and has an arbitrary conduct due to various factors, cannot be changed in a short time. Relevant departments face severe tests in effectively enhancing the capability for enforcing the new *Environmental Protection Law*. The draft amendment to the *Air Pollution Prevention and Control Law* has been finalized in dispute and submitted to the Standing Committee of the National People's Congress for deliberation. The making of the *Soil Pollution Prevention and Control Law* and the *Nuclear Safety Law* is underway.

Keywords

Environmental Protection Law – Air Pollution Prevention and Control Law – Soil Pollution Prevention and Control Law – Nuclear Safety Law

---

* Qie Jianrong, senior journalist of the *Legal Daily*, has been engaged in environmental protection reporting for ten years and has written nearly 3,000 environmental reports, the only winner of the "Trans-Century Cross-China Environmental Protection Tour" Good News First Prize for five consecutive years.

On April 24, 2014, the amendment to the *Environmental Protection Law* was deliberated and adopted during the 8th Session of the Standing Committee of the 12th National People's Congress. The deliberation and adoption of the new *Environmental Protection Law* can be considered a great event in environmental legislation of the year 2014. It stipulates that punishment is imposed on a daily basis, seizure and attachment are made, and administrative detention is enforced against lawbreakers; it specifies the qualifications of the subject for environmental public interest litigation, so it is regarded as the strictest environmental protection law. This law officially came into force as from January 1, 2015.

The future enforcement of this law is not optimistic. There are severe and lingering problems in the real legal environment, laws are not strictly enforced and lawbreakers have not been punished for many years, so it is difficult for a law, even the strictest law, to fundamentally change the situation. Relevant departments face severe tests in effectively improving the capability for enforcing the new *Environmental Protection Law*.

On December 22, 2014, the draft amendment to the *Air Pollution Prevention and Control Law* was submitted to the 12th Session of the Standing Committee of the 12th National People's Congress. Meanwhile, the suggested drafts of the *Soil Pollution Prevention and Control Law* and the *Nuclear Safety Law* have been finalized and submitted to the Environmental Protection and Resources Conservation Committee of the National People's Congress.

1 The Bright Spots in the *New Environmental Protection Law*, such as the "Imposition of a Daily Punishment" and "Public Interest Litigation" Have Attracted Attention

"Imposition of a daily punishment," "seizure and attachment," "administrative detention of lawbreakers" and the mandatory requirement that enterprises and institutions must make environmental information public are undoubtedly the most important part of the newly-amended *Environmental Protection Law*. Therefore, the Ministry of Environmental Protection introduced four supporting measures for the first time, and these measures and the new *Environmental Protection Law* became effective at the same time.

Besides the above characteristics of the new *Environmental Protection Law*, which these provisions constitute, the general public pays more attention to the fact that the new *Environmental Protection Law* dedicates a special chapter to the stipulation of information disclosure and public participation, and in particular, it defines the scope of the subjects for environmental public interest litigations. Both parts of the new *Environmental Protection Law* have the most direct connection with the general public, so they have received the most attention and evaluations.

As we know, environmental protection cannot do without active public participation and supervision. The 5th Chapter of the new *Environmental Protection Law*—Information Disclosure and Public Participation—involves six legal provisions. Making so many legal provisions in order to ascertain the obligations and rights regarding information disclosure and public participation has occurred in China's environmental legislation for the first time.

The chapter "Information Disclosure and Public Participation" provides that citizens, legal persons and non-governmental organizations enjoy the right to obtain environmental information, participate in and supervise environmental protection according to the laws. The major pollutant discharging units must truthfully make public the information concerning the names of their main pollutants, the emission modes, emission concentration and total amount of emissions, emissions beyond the standard, the construction and operation of the pollution prevention and control facilities, so as to accept social supervision. With respect to the construction projects for which an *Environmental Impact Report* has to be prepared according to the laws, the construction unit is obligated to explain, at the time of the preparation of this report, the situation to the people who may be affected so as to fully solicit the public's opinion; the department responsible for examining and approving the environmental impact assessment report on the construction project must, after receiving the *Environmental Impact Report*, release its full text except the matters involving state secrets and business secrets; where the construction project is found to show that no public opinions are fully solicited, the construction unit will be ordered to solicit the public's opinion.

Article 58 of the chapter has an explicit provision for the subjects qualified for filing environmental public interest litigation—the non-governmental organizations which are registered with the civil affairs department of the people's government above the level of the city divided into districts, have specialized in public interest activities relating to the environmental protection for more than five consecutive years and have no records on the violation of laws can bring a lawsuit to the people's court, while the people's court must accept and hear relevant cases according to the laws.

The provisions of the new *Environmental Protection Law* relating to the qualifications of the subjects for environmental public interest litigations cover the shortage in the new *Civil Procedural Law*, effective as from January 1, 2013. The new *Civil Procedural Law* stipulates, at the legal level, that non-governmental organizations can file environmental public interest litigations, but after the enforcement of the new *Civil Procedural Law*, the environmental protection organizations repeatedly met resistance when they filed environmental public interest litigations according to the new law. Before the enforcement of the new *Civil Procedural Law*, some environmental protection organizations, such as the Friends of Nature and All-China Environment Federation, filed some environmental public interest litigations according to some policies and regulations, such as the *Decisions of the State Council Concerning Implementation of the Scientific Outlook on the Development and Reinforcement of Environmental Protection* in 2005; relevant courts accepted and heard relevant cases, and they won lawsuits in not a few cases. However, after the new *Civil Procedural Law* became effective in 2013, no environmental public interest litigation filed by environmental protection organizations has been accepted for hearing by courts. In 2003, the All-China Environment Federation filed 8 environmental public interest litigations, none of which was accepted for hearing by courts. According to the All-China Environment Federation, those were all environmental public interest litigations filed by environmental protection organizations in 2013. After there were legal provisions for environmental public interest litigations, none of them was accepted for hearing in 2013.

The reason why courts refused to accept environmental public interest litigations in 2013 is that though there are legal principles in the new *Civil Procedural Law*, this law does not specify the non-governmental organizations which can file environmental public interest litigations. Therefore, courts cannot judge whether an organization has the necessary qualifications of subject. Courts pointed out that under such a circumstance, courts can only accept no environmental public interest litigation in a unitary way.

Article 58 of the new *Environmental Protection Law* solves the puzzles in the new *Civil Procedural Law*. On the first day of the enforcement of the new *Environmental Protection Law*, the Friends of Nature and another environmental protection organization, the Fujian Green Home, received the notice concerning the acceptance of their environmental public interest litigation case from Nanping Municipal Intermediate People's Court, Fujian Province. On January 13, 2015, two environmental public interest litigation cases instituted by the All-China Environment Federation were successfully filed with Dongying Municipal Intermediate People's Court, Shandong Province.

The new *Environmental Protection Law* dedicates a special chapter to where the information disclosure and public participation are stipulated, and explicitly states that the groups and organizations which meet conditions are qualified for filing public interest litigations. "This is a historic event," said Xia Guang, Director of the Policy Research Center, the Ministry of Environmental

Protection, "the needs of social forces for participating in environmental protection are unprecedentedly clear and strong." Xia Guang believed that the cases of environmental public interest litigation would increase rapidly; though there was not a large number of cases then, many environmental protection organizations were eager to have a try. Xia Guangdong said: "The reason why there is not a large number of cases involving environmental public interest litigations now is that environmental protection organizations need to take time to get familiar with and coordinate with legal provisions."[1]

As the new *Environmental Protection Law* specifies the subject of environmental public interest litigations, the general public is generally optimistic about them and expects that it will serve as a powerful means for public participation and environmental protection.

On January 7, 2015, the supporting document for the new *Environmental Protection Law*—the *Interpretation of Several Issues Concerning the Applicable Laws for Hearing Environmental Civil Public Interest Litigation Cases* was issued. Liao Hong, Deputy Director of the National Non-governmental Organization Administration, the Ministry of Civil Affairs, said that about 700 non-governmental organizations which comply with the *Environmental Protection Law* and its judicial interpretation can file environmental public interest litigations.[2]

If each of these 700 non-governmental organizations files an environmental public interest litigation each year, courts will hear at least 700 cases each year. This is astronomical in the history of environmental public interest litigation in China. Of course, though these environmental protection organizations are qualified for filing environmental public interest litigations, this does not mean that they will certainly bring the public interest litigation cases to court; many organizations are unable to do so due to restrictions from capability, professional knowledge and funds. However, the prospect of environmental public interest litigations is still expectable, and environmental protection organizations will play a big role in them.

Besides these aspects, "imposition of a daily punishment," "seizure and attachment," "administrative detention of lawbreakers" and the mandatory requirement that enterprises and institutions have to make environmental information public are the characteristics of the new *Environmental Protection Law*, considered the strictest one to date. When explaining the new *Environmental*

---

1 Qie Jianrong, "An Expert from the Ministry of Environmental Protection: the Cases of Environmental Public Interest Litigation Will Rapidly Increase," *Legal Daily*. 2015-01-12.
2 Xing Shiwei and Jin Yu, "The Supreme People's Court: 700 Non-governmental Organizations Can File Environmental Public Interest Litigation," *Beijing News*. 2015-01-07.

Protection Law, an official from the Ministry of Environmental Protection pointed out that there was no upper limit for daily punishments and sky-high fines may be imposed. In addition, according to the new *Environmental Protection Law*, administrative detention can be enforced against 23 illegal acts. For the illegal environmental acts, rectification within the specified time limit in the old *Environmental Protection Law* has been replaced by direct seizure and attachment. Severe legal provisions have been available, but uncompromisingly enforcing the law is more important and harder than making the law.

2 It is Difficult to Remove the Deep-rooted Habits that Laws are Not Strictly Enforced and Lawbreakers are Not Punished in a Short Period of Time

It was not easy for the new *Environmental Protection Law*, effective as from January 1, 2015, to be adopted in a new way at the Standing Committee of the National People's Congress. Twenty-five years passed from 1989, when the former *Environmental Protection Law* came into force, until 2014, when it was amended.

The first major amendment to the *Environmental Protection Law* in 25 years resulted from many hardships and struggles among different parties: It spanned two terms of office of the Standing Committee of the National People's Congress, it was deliberated four times, public opinions on it were solicited twice ... Initially, the Environmental Protection and Resources Conservation Committee of the National People's Congress, responsible for the amendment of the law, insisted on minor changes; later, the Legislative Affairs Committee of the Standing Committee of the National People's Congress took charge of the process of amending the law, actively adopted the suggestions from relevant experts and the Ministry of Environmental Protection, and decided to make major amendments, so the 1989 version of the *Environmental Protection Law* was thoroughly remade. Only 2 out of 70 articles in the whole law underwent no changes, while other articles were amended or redrafted. Thanks to a rigorous legislative attitude, the new *Environmental Protection Law* was highly commended immediately after it was adopted. It is generally regarded as the strictest law of its kind by people from all walks of life.

It is not easy to make a law, especially a valuable law, but enforcing a law is harder than making it.

As the most important executor of the new *Environmental Protection Law*, the Ministry of Environmental Protection has fully realized this issue. Both Zhou Shengxian, former Minister of Environmental Protection, and

Pan Yue, vice-minister in charge of the legislation, repeatedly emphasized the importance of the enforcement of the law. On December 30, 2014, the Ministry of Environmental Protection, the Environmental Protection and Resources Conservation Committee of the National People's Congress and the Legislative Affairs Committee of the Standing Committee of the National People's Congress convened a special mobilization meeting for enforcing the new *Environmental Protection Law*, in which Pan Yue stressed that as it was considered the strictest environmental protection law, it would certainly test the capability of the environmental protection department for enforcing it in the strictest way.[3] On January 15, 2015, the National Working Meeting on Environmental Protection was convened, in which Zhou Shengxian, former Minister of Environmental Protection, emphasized that taking the first step in enforcing the new *Environmental Protection Law* was of great importance.

However, since 2015, in particular regarding the implementation of the new *Environmental Protection Law*, fierce violations of it, at some places, have been frequently reported by the media.

On January 15, *Beijing News* reported that many people suffered from cancer due to the pollution from two waste incineration plants at Guodingshan, Yongfeng Village, Hanyang District, Wuhan City. On the same day, the *Legal Daily* reported that the Jinnan Paper Mill in Qingyuan County, Baoding City, Hebei Province had created a false impression before the personnel of the environmental protection regulation at the provincial, municipal and county levels that the sewage treatment plant was under operation, and provided false information on the COD online monitoring, and villagers jointly appealed to the higher authorities for help but received no response. On January 16, the CCTV reported that there were many cases of cancer in Shenzhou County, Hebei Province due to long-term pollution from pharmaceutical factories.

There is a view that violations of the environmental protection law will dramatically decline after the enforcement of the new *Environmental Protection Law*. However, actually, it is far from being so optimistic. This is because it is unrealistic to thoroughly remove, in a short time, the long-standing problems that laws are not strictly enforced and lawbreakers had not been duly punished in the original legal system.

Zhou Shengxian, former Minister of Environmental Protection, has analyzed the die-hard bad habits in environmental supervision and environmental law enforcement for a long time. He believed that some typical cases have exposed some conspicuous problems that "violations of laws will not be

---

3 Qie Jianrong, "The Strictest Environmental Protection Law Is Expected to Be Enforced in the Strictest Way," *Legal Daily*. 2015-01-05.

investigated," "violations of laws cannot be investigated," "inaction occurs" and "there is some arbitrary conduct" in the enforcement of environmental laws.[4] Frequent occurrences of cases of violation of environmental laws have been closely related to the above problems arising out of some grass-roots environmental law enforcement personnel; however, the above problems result from long-term accumulation and are closely connected to the low quality of the professionalism of grass-roots environmental law enforcement personnel, and their long-term slackness in law enforcement, even their collusion with law-breaking enterprises. The above problems have not been corrected for a long time, so they have become increasingly salient.

The government department for environmental protection has relatively high requirements regarding the occupational level; however, local environmental protection authorities, especially those at the county level, often become the "idle" departments where some leaders arbitrarily arrange their friends and relatives with jobs. Some people who know little about environmental protection and may not even understand the operation of sewage treatment plants and online monitoring instruments, are assigned to environmental protection authorities and sent to the front line of law enforcement. These problems are widespread within the grass-roots environmental protection departments.

Sun Youhai, former Director of the Office of Proposed Laws, the Environmental Protection and Resources Conservation Committee of the National People's Congress, current Director of the Applied Jurisprudence Research Institute, Supreme People's Court, said that the environmental behavior consistent with environmental laws and regulations accounted for only 30%, and about 70% of the environmental laws and regulations were not observed in China.[5]

The new *Environmental Protection Law* incorporates the strictest articles, but in reality, no new grass-roots environmental law enforcement teams for enforcing the new *Environmental Protection Law* have been established, so there is a change in form, but not in content. Therefore, from this perspective, the prospects for enforcing the new *Environmental Protection Law* are not very optimistic. Can the new *Environmental Protection Law*—the

---

4 Qie Jianrong and Zhou Shengxian, "Called for Better Taking the First Step to Enforce the New Environmental Protection Law," *Legal Daily*. 2015-01-16.
5 Qie Jianrong, "We Hope That the Environmental Protection Law Can Contain Illegal Acts, Making Public Pollution Information Is More Important Than Imposing Fines," *Legal Daily*. 2013-07-31.

strictest environmental protection law—break the curse that laws are not strictly enforced and lawbreakers are not punished? The answer remains unknown.

3 The Draft Amendment to the *Air Pollution Prevention and Control Law* is Controversial

China's *Air Pollution Prevention and Control Law* was made in 1987, then amended in 1995 and 2000. Currently, this law is undergoing its first major amendment in 15 years.

This amendment to the *Air Pollution Prevention and Control Law* was initiated as early as 2006. The draft amendment was submitted to the Legislative Affairs Office of the State Council in 2010; afterwards, it was shelved for a long time. The increasingly severe haze pollution which has frequently occurred in recent years has hindered the progress in amending the *Air Pollution Prevention and Control Law* on the one hand, and has accelerated the amendment to this law on the other hand.

The reason why the amendment to this law was shelved is that the haze has actually worsened in recent years; previous amendments did not give any more consideration to haze, but now haze pollution control must be regarded as an important aspect. It must be incorporated into the amended *Air Pollution Prevention and Control Law*. Furthermore, the State Council released the *Action Plan for Air Pollution Prevention and Control* in 2014, so the amendment to the *Air Pollution Prevention and Control Law* shall fix some provisions of this action plan in the form of laws. Given both reasons, the *Air Pollution Prevention and Control Law* must be amended. In this sense, haze has, to some extent, affected the progress in law amendment. Moreover, the severe haze pollution has expedited the introduction of the *Air Pollution Prevention and Control Law*; from this perspective, haze pollution has speeded up the amendment of the law.

With repeated feasibility studies for more than 8 years, the draft amendment to the *Air Pollution Prevention and Control Law* was submitted to the 12th Session of the Standing Committee of the 12th National People's Congress for deliberation on December 22, 2014. This is the third amendment in 27 years since the enforcement of the *Air Pollution Prevention and Control Law* and the greatest amendment, but there is still a view that even if the amendment to the *Air Pollution Prevention and Control Law* were adopted, it might fail to completely solve the problem of air pollution. Such a view holds that making and introducing the *Clean Air Law* is the fundamental policy.

4 The *Soil Pollution Prevention and Control Law* and the *Nuclear Safety Law* were Being Drawn Up

So far, China has made more than 30 environmental protection laws, including the *Environmental Protection Law*, the *Marine Environmental Protection Law* and the *Water Pollution Prevention and Control Law*. There are more than 90 administrative regulations, including the *Administrative Regulations for the Collection and Use of Pollution Discharge Fees* and the *Administrative Regulations for Environmental Protection of Construction Projects*. Therefore, about 120 environmental protection laws and regulations are available in China.

Although environmental legislation tops the list of legislation in various fields, some basic laws are still unavailable. The most typical ones among them are the *Soil Pollution Prevention and Control Law* and the *Nuclear Safety Law*.

In recent years, China's land pollution problem has become increasingly severe. According to the *Communiqué of the National Survey on the Status of Soil Pollution* jointly released by the Ministry of Environmental Protection and the Ministry of Land and Resources on April 17, 2014, the overall situation of the national soil environment was not optimistic, soil pollution in some areas was relatively severe, the quality of the soil environment of the cultivated land was worrying, and the problems concerning the soil environment of the industrial and mineral wasteland were prominent. The overall national rate of point locations exceeding the standard was 16.1%—the slight, mild, moderate and heavy pollution point locations accounted for 11.2%, 2.3%, 1.5% and 1.1% respectively.

Soil pollution is becoming increasingly severe, but it is not governed by special laws. Amidst repeated calls, the proposed draft of the *Soil Pollution Prevention and Control Law* was developed and submitted to the Environmental Protection and Resources Conservation Committee of the National People's Congress in 2014.

As of 2014, China's nuclear industry had existed for 60 years and the civil nuclear energy had been developed and utilized for 30 years. As indicated by the statistical data from the Ministry of Environmental Protection, with development for many years, China ranks No.1 in the world in terms of the number of nuclear power units under construction. According to the *Medium and Long-term Development Plan for Nuclear Power* approved by the State Council in 2012, by 2020, China's installed capacity of nuclear power will reach 88 million kW, ranking No. 2 in the world, second only to the USA.

China has become a real great world power in the area of nuclear energy, but the fundamental law in the area of nuclear safety—the *Nuclear Safety Law*—is not available. In 2014, new progress was made in making this law. According to the Ministry of Environmental Protection, the proposed draft of the *Nuclear Safety Law* was developed and submitted to the Environmental Protection and Resources Conservation Committee of the National People's Congress in 2014.

CHAPTER 7

The Financial Sector and Environmental Risks: Understanding the New *Environmental Protection Law*

WANG Xiaojiang and WANG Tianju*

Abstract

The new *Environmental Protection Law* became effective as of January 1, 2015. For the financial macro-control department and commercial financial institutions, some changes have taken place in the ecology of finance. The new *Environmental Protection Law* will change the original rules of operations and the survival condition of enterprises with respect to environmental protection, and will generate many new operational risks, thus affecting the changes in the ecology of finance and financial risks. If financial regulators do not effectively control and deal with them, it is very likely that they will become systematic financial risks, hindering and threatening the operational safety of the financial system and financial institutions. Against such a background, the construction of a legal system, institutional system, responsibility system and an organizational management system for guarding against and controlling financial environmental risks and the cultivation of independent social third-party forces will become an important task for the financial regulators and financial institutions in a certain period of time in the future.

---

* Wang Xiaojiang, Director, Professor at the Green Financial Research Institute, Hebei University of Economics and Business, mainly engaged in research on financial management, project evaluation, risk management, compensation for damages, equity and debt operation practice, financial product design, performance management and evaluation of funds. He cooperates with a number of international organizations and participates in research projects concerning green finance, taking charge of and participating in the national institutional construction relating to green finance.
   Wang Tianju, Project Specialist at Greenovation Hub, has been engaged in environmental and financial research. He has a Bachelor of financial economics from Swansea University, UK and a Master's in climate change and risk management from the University of Exeter, UK. He focuses on the issues concerning green finance, conducts research on climate risk identification, analysis and management at financial institutions, and participated in designing and researching the issues concerning the implementation of local green credit policy during the period of his internship.

Keywords

new *Environmental Protection Law* – environmental risks for enterprises – financial environmental risks – green finance

The overall objective of the new *Environmental Protection Law* to be enforced in 2015 is to safeguard balanced social interests and long-term sustainable development, and to ensure ecological safety and environmental health. Its core lies in imposing strong constraints on the environmental behaviors of society and enterprises so as to turn the predatory and extensive economic mode into a harmonious economic model, and provide the general public with good basic living conditions.

With regard to the investing and financing behaviors of financial institutions, the environmental risks for enterprises involve many hidden, cumulative, long-term, accidental and devastating disputes, high compensation and a long time, so the changes and trends of environmental protection laws, regulations and national policies will exert a great long-term impact. In the long term, the new *Environmental Protection Law* is conducive to preserving financial order, promoting rational allocation of financial resources and a good cooperative relationship between financial institutions and enterprises; however, in the short term, the financial regulators and financial institutions have not yet made the corresponding arrangements for the introduction of the new *Environmental Protection Law*, so it is difficult to adapt to the basic requirements of this new law and the changes in the operating environment for enterprises in a short term, and risks may sharply increase in a certain period of time.

1 New Risks Which the New *Environmental Protection Law* May Present to Financial Institutions

1.1 *The Risk of Environmental Responsibility for Enterprise Owners and Operators as Individuals*

The center of an enterprise is people, while the entrepreneur is the center of enterprise operations. Once the core people at enterprises as debtors become problematic, the operational prospects will certainly look bleak; this is the greatest risk for financial institutions engaged in investing and financing. The new *Environmental Protection Law* grants a more regulatory power to governments at various levels and environmental protection departments, and permits the adoption of multiple new means to exercise such power. It is noteworthy that it also uncommonly specifies the punitive measures of administrative detention, and permits the adoption of the severest means of administrative punishment against the environmental law-breakers. Moreover, it makes it possible for the environmental regulator to get access to the site for inspection, order the pollutant discharging units with discharge exceeding the standard and total quantity so as to limit or stop production. If relevant enterprises do not control pollution, the people in charge of the law-breaking enterprises which carry out construction before approval and refuse to make corrections, and discharge pollutants through concealed pipes to dodge regulations will be subject to detention based on the new *Environmental Protection Law*; where they commit crimes, they will be held accountable for criminal liability according to laws. In the future, the entrepreneurs and senior executives may face more kinds of punishment for environmental responsibility, which will directly affect the safety of investments from financial institutions.

1.2 The Risk is that Enterprises Might Be Subject to Continuous, Exceedingly Large Economic Punishments

According to Article 59 of the new *Environmental Protection Law*, where enterprises, public institutions and other production operators discharge pollutants and in so doing violate laws, they will be subject to penalties and ordered to make corrections; if they refuse to make corrections, the administrative organ which makes the penalty decision according to laws can consecutively punish them on a daily basis according to the original amount of penalty as of the day after the day when they are ordered to make corrections.[1] With respect to such a principle of law, the concurrently enforced *Measures by the Competent Department for Environmental Protection for Imposing Consecutive Daily Punishment* explicitly stipulates that five illegal pollutant discharging behaviors are subject to daily punishment; the five illegal behaviors include: exceeding the national or local pollutant discharge standard, or discharging pollutants beyond the indicators of the total quantity control of major pollutants; discharging pollutants by means of concealed pipes, seepage wells, seepage pits, filling or tampering with, or falsifying monitoring data, or abnormally operating pollution prevention and control facilities in order to dodge the regulation; discharging the pollutants which are prohibited from being discharged by laws and regulations; dumping hazardous wastes in violation of laws; other behaviors of illegally discharging pollutants. The "daily punishment" system

---

1 *Environmental Protection Law of the People's Republic of China.* http://rsj.huizhou.gov.cn/publicfiles/business/htmlfiles/1274/2.1/201501/352859.html.

is designed to punish the behaviors of the subjectively malicious violation of environmental laws, while its most important part consists in urging lawbreaking enterprises to quickly stop their illegal behaviors. If enterprises fail to stop them promptly, penalties will be imposed until production and operating activities become unsustainable.

The above provisions have a great risk impact on the investments from financial institutions—first, because punishment for environmental responsibility is imposed, economic benefits for enterprises drastically decline, the normal operation is affected and the risk for the payment of interests from enterprises increases, thus lowering the benefits for financial institutions; second, enterprises suffer losses due to penalties, in which case, the possibility of interest payment decreases, there are more difficulties for enterprises to repay the principal, loss of principal may occur, directly affecting the recovery of investments for the financial institutions and then the normal operations of financial institutions.

1.3 The Risk that Enterprises Will Have to Pay Huge Amounts of Compensation and Remediation Expenses

The new *Environmental Protection Law* establishes the basic principle that those who/which cause damage shall assume liability—those who/which pollute the environment, destroy the ecology, harm or injure people or livestock shall assume liability. "Assuming liability" means assuming the liability for making compensation, restoring the environment, remediating the ecology or paying the above expenses, while "damage" means the behaviors which exert any adverse environmental impact, including those which degenerate the capability for self-restoration on the part of the environment due to the bad utilization of the environment. The principle that "those who/which cause damage shall assume liability" means that the people who exert any adverse environmental impact shall undertake the legal duty or legal liability for restoring the environment, remediating the ecology or paying the above expenses.

The new *Environmental Protection Law*, in particular, authorizes the nongovernmental organizations which satisfy the conditions to file environmental public interest litigation, and expands the subjects of environmental public interest litigation, providing the basic guarantee for the assumption of environmental liability. Predictably, it is altogether possible that increasing public interest and private interest litigations will make enterprises become subject to compensation and joint liability after they lose litigations.

Huge compensation made by enterprises presents dual risks to financial institutions. For enterprises as debtors, if the amount of the environmental responsibility compensation and the subsequent remediation falls within the asset capability of the enterprises, the financial institutions shall bear the risk of delaying fund repayment, and such a risk is controllable; however, if the amount of the environmental responsibility compensation and the subsequent remediation goes beyond the fund capability of the enterprises, the enterprises may enter a bankruptcy procedure, and risks will further increase, directly threatening the principal safety of the financial institutions.

1.4 The Risk that Enterprises Might Be Restricted from Carrying out Production

According to the new *Environmental Protection Law*, where pollutant discharging units exceed the pollutant discharge standard or the daily maximum quantity of discharge allowable of major pollutants, the competent department for environmental protection may order them to adopt measures to cease production. If the financial institutions fail to effectively match the maximum quantity of allowable discharge of the enterprises with the loan limit for the enterprises before extending the loans, and increase the financing limit for the enterprises, the financial institutions will bear the risk loss incurred.

1.5 The Risk that Enterprises are Ordered to Stop Production

According to the new *Environmental Protection Law*, under one of the following circumstances, the competent department for environmental protection will, through the people's government with the authority for approval, order the pollutant discharging units to stop doing business, or to shut down: first, the pollutant discharging units undergo an administrative penalty more than twice in two years because the toxic substances containing heavy metals and persistent organic pollutants from them exceeded the pollutant discharge standard, and once again commit the above act; second, the pollutant discharging units refuse to stop production or restore production without authorization after they are ordered to stop production for rectification; third, after the decision of stopping production for rectification is removed, a follow-up inspection shows that they have committed the same illegal act again; fourth, other circumstances under which environmental laws and regulations are severely violated. These four circumstances share a common characteristic: the circumstance is serious as provided in Article 60 of the new *Environmental Protection Law*. Shutting down business means that enterprises lose their valuable creative ability and various resources are left unused, so it is very disruptive to enterprises and incurs huge financial losses to the financial institutions. Therefore, carrying out surveys, evaluation and forecasts of these environmental risks is the priority of the risk management performed by the relevant personnel of the financial institutions.

1.6 The Risk Involving Enterprises' Information Disclosure and Public Participation

According to Article 53 of the new *Environmental Protection Law*, citizens, legal persons and other organizations enjoy the right to obtain environmental information, participate in and supervise environmental protection according to laws. The competent departments for environmental protection and other departments for supervision and management of environmental protection of the governments at various levels shall make environmental information public according to laws, improve the public participation procedure, and provide citizens, legal persons and other organizations with the convenience of participating in and supervising environmental protection.

The new *Environmental Protection Law* also expands the scope of the subjects of environmental public interest litigation, and stipulates that the non-governmental organizations which are registered with the civil affairs department of the people's government above the level of the city divided into districts, have specialized in public interest activities relating to environmental protection for more than five consecutive years and enjoy a good reputation can file a litigation with the people's court. This is of great significance for and plays an important role in enhancing public awareness about environmental protection, developing the philosophy of public participation in this issue, promptly discovering and stopping illegal environmental behaviors.

With regard to enterprise information disclosure and environmental impact assessment, it is necessary to specify the subjects of the public participation and environmental public interest litigation so as to further boost the interaction among the environmental behaviors of the enterprises, the surrounding communities and the general public. The interest appeals of the surrounding communities and the general public will further influence the development and operations of enterprises, and may even become the decisive factor under certain conditions.

Enterprise information disclosure and the determination of the subjects of public participation and environmental public interest litigation are a double-edged sword for financial institutions—on the one hand, it is simpler and more periodical for the financial institutions to obtain the environmental information of the enterprises as debtors, and the survey, analysis and evaluation conducted by the financial institutions become more accurate; on the other hand, the general public learns about the situation of the operation of the enterprises and can, where the environmental behaviors of enterprises are abnormal, make a quick response to safeguard their rights by means of lodging complaints and filing litigations. Therefore, the financial institutions must attach importance to the opinions of the surrounding communities and the general public with respect to their investment behaviors, change the original risk control mode and standard, and establish a new financial risk control system.

The above six environmental risks are discussed as single risks under certain conditions, but in reality, environmental risks may often occur in a cascading way—comprehensive environmental risks may pop up; in other words, the above environmental risks may arise in a cascading way in a certain period of time. They crop up in the following two ways: First, various risks involving the environmental behaviors of enterprises occur in a comprehensive way, in which case, these risks are only limited to single enterprises, and the main measures adopted for risk control are concentrated in single enterprises, so these risks are controllable in principle; second, the environmental risks intensively loom large within a certain scope, and this way is extremely critical for the financial institutions; furthermore, in this case, as illegal environmental behaviors of the enterprises and environmental credit risk for financial institutions exist side by side, financial risks may consequently occur on a larger scale. The environmental regulator and the financial regulators must guard against and avoid such a situation as far as possible.

2 Construction of New Financial Management Systems Adaptive to Environmental Risk Changes

The environmental risks are more destructive than generic risks involving the operating behaviors of enterprises; they result from the multi-party interest game among the society, the environment and enterprises, and may even incur certain political risks. Therefore, the environmental risks should not be managed and controlled by merely dealing with the relationship and interaction between the financial institutions and enterprises; on the contrary, it is essential to take the perspective of national financial security, environmental and social safety to understand the prevention of the environmental risks and the construction of management systems.

The environmental risks come from enterprises and are caused by the externality of the enterprises' environmental behaviors, while the management of environmental risks involves the government and the general public. The management of the financial environmental risks relates to financial regulators and financial enterprises. Environmental risk management involves different fields and industries, and is cumulative, concealed and long-term, so it is a systematic project and a complete construction project with its own philosophy, behaviors and management systems.

Thinking serves as a guide to one's actions, while the degree of financial environmental risk control depends on the depth of thought and the extent to which the particular line of thought is accepted and popularized. Therefore, first, it is necessary to develop the concept of financial environmental risk in the whole society, make people participate in environmental risk control and foster a social atmosphere of green finance. Second, the awareness of green finance and environmental risks should be developed within the financial industry, and environmental risk awareness should be incorporated into industrial norms, business comparison and appraisal, as well as credit evaluation. Third, environmental risk awareness should really be carried out in every work process of financial institutions, and cover the whole process including strategy making, market positioning, information survey, risk evaluation, investment decision-making, tracking and early warning, work reporting and cost analysis. The line of thought should be closely and completely combined with work practice.

The construction of new management systems is the basic premise and task for managing environmental risks. According to the work principle and requirements for the financial management of environmental risks, financial management systems for coping with environmental risks can be built according to the following three levels: First, actions are taken to build the national financial management systems for coping with environmental risks, including the macro environmental risk management system of the People's Bank of China, the environmental risk management system of the banking regulator, the environmental risk management system for the insurance industry, the environmental risk management system for the securities industry and the environmental risk management systems of other financial regulators; second, the environmental risk management system of the financial industry is established—an environmental risk management mechanism with self-supervision, self-disclosure and self-discipline is developed within the financial industry to achieve a fair mechanism within the environmental risk management industry; third, the financial institutions should build a working system and mechanism for environmental risk management to assign the environmental risk management work to departments, posts and the people in charge of the work.

3 Construction of a New Financial Institutional System that is Adaptable to Environmental Risk Changes

3.1 *Construction of the Legal System for Financial Environmental Risks*

The prevention and management of financial environmental risks calls for a strong guarantee from a solid legal and regulatory system, especially addressing the real problems and defects in China's financial field. Currently, the laws and regulations for financial environmental risks are not available in China; as a result, the responsibilities in the relevant management process cannot be carried out, and environmental risk management is unclear, uncoordinated and not unified. Given the current situation of China's environmental protection and China's strategic goal for the construction of an ecological civilization, there is an urgent need for making the legal provisions in coordination with the financial environmental risk system, and ensuring that laws and regulations are more operable, and the green financial system is effectively implemented. Meanwhile, the laws and regulations for financial environmental risks should be taken as the yardstick, and the capability of the law enforcement system should be enhanced during the implementation of the financial environmental risk system.

3.2 Construction of the Institutional Systems for Financial Environmental Risk Control

In order to achieve the goal of financial environmental risk management, proceeding from China's national conditions and drawing upon experience from various other countries is a priority for building an institutional system. The most important systems are the following basic ones having a bearing on the overall situation: First, the standard system for financial environmental risks; second, the financial environmental risk evaluation and monitoring system; third, the financial environmental risk reporting system; fourth, the financial environmental risk prevention management system; fifth, the financial environmental risk goal responsibility, reward and punishment system; sixth, the financial environmental risk post responsibility system; seventh, the credit management system for financial environmental risks; eighth, the financial environmental risk compensation system, etc.

3.3 Construction of the Financial Environmental Risk Responsibility System

The implementation of responsibilities is the foundation for environmental risk management and the core in building the environmental risk management mechanism. Therefore, the construction of the financial environmental risk responsibility system is the central task in financial environmental risk management. The responsibility system mainly covers: First, the construction of the subjects of responsibility at three levels—the environmental risk management responsibility of the environmental regulators, the environmental risk supervision and management responsibility of the financial regulators (People's Bank of China, China Securities Regulatory Commission, China

Banking Regulatory Commission, China Insurance Regulatory Commission and other financial regulators), and the environmental risk investigation and evaluation responsibility of various commercial financial institutions; second, the construction of the financial goal management system for environmental risks; third, the construction of the financial management mechanism for environmental risks; fourth, the construction of the performance management evaluation mechanism for environmental risks, etc.

4   Construction of the New Management Systems of the Financial Institutions Able to Adapt to Environmental Risk Changes

Amid the changes in the situation of environmental risk, carrying out preventive and control measures will be the priority and direction for the work of financial institutions in environmental risk control. Specifically, the following aspects are covered: First, the implementation of a relevant strategy—the awareness about environmental risk prevention and its relevant work content are incorporated into the risk management strategy of the financial institutions, and a risk strategy of the financial institutions consistent with their strategies is developed; second, the implementation of the responsibility for the environmental risk work—a specialized environmental risk management body is established to be responsible for management and coordination, and professional talents for environmental risk management are chosen and responsibilities are assigned to individuals; third, the work on environmental risk prevention is incorporated into the whole process of investment analysis, and environmental risk management is integrated into every work process, from enterprise survey, credit evaluation, investment decision-making, post-supervision to work reporting; fourth, an environmental risk reporting system, covering enterprise environmental risk reports and environmental risk reports of the financial institutions, is built to promptly handle environmental risk information and form a complete environmental risk information management system.

5   Cultivation of Independent Social Third-Party Forces Able to Adapt to Environmental Risk Changes

As shown by the experience and practice of various countries in the world, the important prerequisite for solving environmental problems is information disclosure and the involvement of independent third-party bodies. That involvement is an effective solution to ensure that the real multi-party information concerning the environmental behaviors of the enterprises is disclosed.

5.1 Build a Social Participation Mechanism for Environmental Risk Management

It is necessary to develop a working platform for effective communication, invite social volunteers in environmental protection, environmental protection organizations, professional bodies and relevant government departments to effectively interact with the financial institutions in every process, from approval and initiation of major environmental protection projects to their evaluation, review, decision-making and implementation, and to participate in the credit decision-making of the financial institutions. On the one hand, this makes it possible to avoid environmental risks for the financial institutions; on the other hand, it helps enable benign interaction between financial investment behaviors and society.

5.2 Establish a Cross-Institution and Trans-Department Environmental Risk Information Exchange Platform

Information is the core in the decision-making of financial institutions. The work of the green financial information exchange platform covers governmental environmental protection information, enterprise environmental behavior information, credit information of the financial institutions, decision-making information of the financial institutions, information on post-loan environmental behaviors of enterprises and information on post-loan evaluation of the financial institutions. Financial institutions, environmental protection departments and third-party evaluation bodies interact with each other on this platform.

5.3 Set Up Independent Third-Party Green Credit Evaluation Bodies

Credit evaluation is a long-acting management mechanism for environmental risks. First, involvement in the capacity of the independent third party can ensure the independence of behaviors and effectively prevent the personnel within the financial institutions from being driven by interests to make directional mistakes in their behaviors; second, the selection of an independent third party ensures that the environmental risk management technologies are professional and provides the basic technical support for environmental risk control.

Establishing a new financial management mode for environmental risks will be an important task for the financial regulators and financial institutions in the future. It should be integrated into various aspects and the whole process of financial laws, financial system, green credit evaluation and the construction of a financial culture, so that financial behaviors can promote sustainable development of the enterprises and society.

CHAPTER 8

Recent Developments in Environmental Criminal Justice

*Yu Haisong and Ma Jian**

Abstract

Based on the cases involving environmental pollution crimes[1] heard by the people's courts from July 2013 to October 2014, this chapter presents the characteristics of the implementation of the *environmental judicial interpretation of the Supreme People's Court and the Supreme People's Procuratorate* since the promulgation of this environmental judicial interpretation. They have the following characteristics: (1) The criminal cases involving environmental pollution dramatically increased; (2) The criminal cases involving environmental pollution featured an uneven geographical distribution; (3) The criminal cases involving environmental pollution were extensively governed by the Interpretation; (4) The criminalized cases due to environmental pollution fell outside the group of enterprises above the designated scale; (5) The hazardous waste criminals were severely punished; (6) The crime of illegally disposing of imported solid wastes was activated; (7) The criminal cases involving dereliction of duty in environmental regulation markedly declined. According to an analysis of the above characteristics, the authors put forward several issues concerning urgent reinforcement and improvement in environmental judicial practice.

Keywords

environmental judicial interpretation of the Supreme People's Court and the Supreme People's Procuratorate – implementation overview – implementation characteristics

---

*   Yu Haisong, judge at the Office of the Criminal Department of Research, the Supreme People's Court, Doctor of Judicial Science; Ma Jian, cadre at the Statistics Office of the Research Office, the Supreme People's Court, Master of Law.
1   Environmental pollution crimes include four crimes as specified in the Interpretation—the environmental pollution crime provided in Article 338 of the *Criminal Law*, the crime of illegally disposing of imported solid wastes and the crime of importing solid wastes without authorization provided in Article 339, and the crime of dereliction of duty in environmental regulation provided in Article 408.

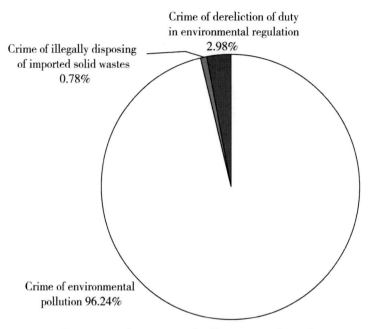

FIGURE 8.1 Composition of environmental pollution crimes from July 2013 to October 2014

As the severest judicial interpretation for environmental protection, the *Interpretation of the Supreme People's Court and the Supreme People's Procuratorate of Several Issues Concerning Applicable Laws for Handling Criminal Cases Involving Environmental Pollution* (Fa Shi [2013] No. 15, "the Interpretation") has drawn extensive attention since its implementation on June 19, 2013. It exerts an important impact on the environmental administrative law enforcement and criminal justice. Based on the criminal cases involving environmental pollution heard by the people's courts from July 2013 to October 2014 (see Figure 8.1), this chapter summarizes and analyzes the characteristics of the implementation of the Interpretation, and presents the suggestions for further improving environmental criminal justice.

1 Overview of the Implementation of the Interpretation

The people's courts, the people's procuratorates, the public security organs and environmental protection departments at various levels took the opportunity of the promulgation and implementation of the Interpretation to continuously keep a tough stance on criminal behavior involving environmental pollution, ascertain facts, correctly apply laws, and resolutely punish the criminal activities relating to environmental pollution according to the laws, thus producing a positive social effect. From July, 2013 to October, 2014, the courts across China received 1,025 criminal cases involving environmental pollution, illegal disposal of imported solid wastes and dereliction of duty in environmental regulation, heard and closed 772 cases, and entered effective judgments against 1,136 people[2]—991 criminal cases involving environmental pollution were received, 743 cases were heard and closed, effective judgments were entered against 1,104 people; 6 criminal cases involving illegal disposal of imported solid wastes were received, 6 cases were heard and closed, effective judgments were entered against 6 people; 28 criminal cases involving dereliction of duty in environmental regulation were received, 23 cases were heard and closed, effective judgments were entered against 26 people.

2   Characteristics of the Implementation of the Interpretation

2.1   *Criminal Cases Involving Environmental Pollution Surged*

After the Interpretation was implemented, criminal cases involving environmental pollution markedly increased. The number of cases governed by Article 338 of the revised *Criminal Law*[3] has risen from 1 digit to 2, 3 and 4 digits since the *Criminal Law* revised in 1997 became effective. In general, the number of cases governed by Article 338 of the revised *Criminal Law* was a 1-digit number and did not exceed 10 before 2006; the number of cases governed by Article 338 of the revised *Criminal Law* was about 20–2 digits—from 2007 to 2012; the number of cases governed by Article 338 of the revised *Criminal Law* was 104 and reached 3 digits for the first time in 2013; the number of cases governed by Article 338 of the revised *Criminal Law* was 656 during the period January–October, 2014, given that 223 cases were newly received but had not yet been heard and closed, the number of cases heard and closed in 2014 reached 4 digits (see Figure 8.2).[4]

---

2   During this period, the people's courts did not hear criminal cases involving the importation of solid wastes without authorization.
3   Article 338 of the *Criminal Law* revised in 1997 defines the crime of a major accident regarding environmental pollution; the *Eighth Amendment to Criminal Law* effective on May 1, 2011 revises Article 338 of the *Criminal Law* by adjusting the crime as the crime of environmental pollution.
4   The people's courts received 879 criminal cases involving environmental pollution during January–October, 2014.

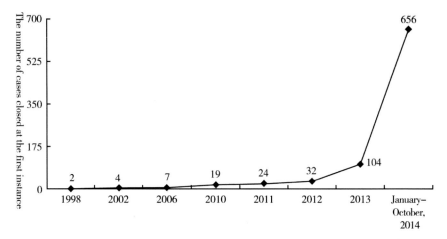

FIGURE 8.2 Changes regarding the increase in criminal cases involving environmental pollution

A sharp increase in the criminal cases involving environmental pollution was due to the fact that the parties from various sectors of the society had fundamentally changed their tolerance towards environmental pollution, and the people in the society had reached a consensus on intensifying punishment of crimes regarding environmental pollution, thus practically reversing the situation where importance was attached to economic development but less attention was paid to environmental protection. The specific reasons can be summarized as follows: First, the Interpretation specifies the criterion for the conviction and sentencing concerning crimes of environmental pollution, overcomes the difficulties in obtaining evidence, authenticating and ascertaining in practice in a well-targeted manner, and provides an effective legal weapon for investigating, transferring and hearing the criminal cases involving environmental pollution. In particular, based on the location of the pollutant discharge, the quantity of that discharge, the degree of discharge, the manner of the discharge and the criminal record of those involved, Paragraph 1–Paragraph 5, Article 1 of the Interpretation adds several specific criteria for ascertaining severe environmental pollution to criminalize the behavior of polluting the environment. These several criminalization criteria were crucial for the increase of the number of cases. Second, environmental protection departments intensified their efforts to investigate and transfer the criminal cases involving environmental pollution. According to relevant statistical data, environmental protection departments transferred nearly 372 cases involving environmental pollution to public security organs in 2013, which were the total of those in the past ten years.⁵ Third, the public security organs unprecedentedly cracked down on the crimes of environmental pollution. The public security organs registered and investigated 779 cases and captured 1,265 criminal suspects—the public security organs took the initiative to investigate 407 out of these 779 cases.⁶

2.2 Criminal Cases Involving Environmental Pollution were Unevenly Distributed from a Geographical Perspective

A total of 991 cases have been received by the courts nationwide, and 743 criminal cases involving environmental pollution have been heard and closed since the Interpretation was implemented; the geographical distribution of these cases was obviously uneven—the cases received and closed in Zhejiang Province made up 50% of those across China, 468 cases were received and 391 cases were heard and closed in Zhejiang Province, accounting for 47% and 53%, respectively. Moreover, more than 80% of the criminal cases involving environmental pollution across China were concentrated in four provinces and one municipality, which are Zhejiang, Hebei, Shandong, Tianjin and Guangdong. A total of 809 cases were received and 610 were closed in these five places, accounting for 82% and 82% of total cases received and closed nationwide, respectively—154 cases were received, 74 cases were heard and closed in Hebei, accounting for 16% and 10%, respectively; 68 cases were received, 52 cases were heard and closed in Shandong, accounting for 7% and 7%, respectively; 59 cases were received, 51 cases were heard and closed in Tianjin, accounting for 6% and 7%, respectively; 60 cases were received, 42 cases were heard and closed in Guangdong, accounting for 6% and 6%, respectively (see Figure 8.3).⁷ On the contrary, there were few provinces, autonomous regions and municipalities directly under the Central Government where no criminal cases involving environmental pollution had been heard since the Interpretation was implemented.

The uneven geographical distribution of criminal cases involving environmental pollution occurred for very complicated reasons. However, the number

---

5 See the exclusive interview with Huang Ming, Vice-minister of the Ministry of Public Security, at the CCVT program *Xiaosa's Interview in Two Sessions* in March, 2014.
6 See the exclusive interview with Huang Ming, Vice-minister of the Ministry of Public Security, at the CCVT program *Xiaosa's Interview in Two Sessions* in March, 2014.
7 In addition, 52 cases were received, 27 cases were heard and closed in Fujian; 40 cases were received, 30 cases were heard and closed in Jiangsu; 21 cases were received, 18 cases were heard and closed in Shanghai; 14 cases were received, 11 cases were heard and closed in Anhui; 11 cases were received, 9 cases were heard and closed in Henan.

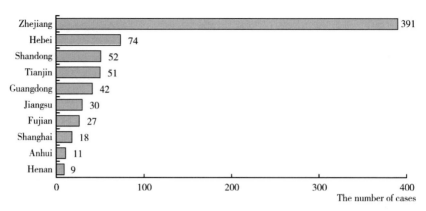

FIGURE 8.3 Distribution of cases closed in the first instance from July 2013 to October 2014

of cases was not necessarily associated with the degree of local environmental pollution, but was closely related to the degree of local attention paid to the problems regarding environmental pollution. There were more criminal cases involving environmental pollution heard in Zhejiang mainly because of the high attention paid to the problem at the local level—Zhejiang Provincial Party Committee and Provincial Government put forward the major policy of treating sewage, preventing flooding, discharging stagnant water, guaranteeing the water supply and saving water. Under the guidance of this policy, the environmental protection departments and the public security departments of Zhejiang Province stepped up their efforts to investigate the crimes of polluting the environment, especially water bodies, so the cases accepted for hearing by courts evidently increased.[8] Take Wenzhou of Zhejiang as an example, the environmental protection departments transferred 240 environmental criminal cases to the public security organs, and the public security organs detained 391 people in a year after the implementation of the Interpretation, from July 2013 to June 2014.[9] The fact that there were a large number of criminal cases involving environmental pollution in Hebei was closely related to the strict investigation of criminal cases involving environmental pollution in Hebei; in particular, Hebei became the first area in China to establish environmental safety brigades which have played an important role in cracking down on the crimes of environmental pollution.

---

[8] See the speech delivered by Chen Guangduo, Presiding Judge of No.1 Criminal Tribunal, Zhejiang Provincial Higher People's Court, at the news briefing "Zhejiang's Courts Act on Treating Sewage, Preventing Flooding, Discharging Stagnant Water, Guaranteeing the Water Supply and Saving Water to Build a Beautiful Zhejiang and Create a Good Life According to Laws" in June 10, 2014.

[9] "All-out Efforts Were Made in Law Enforcement, No Law-breakers Can Evade Legal Sanctions," *China Environment News*. 2014-07-23, page 1.

The authorities in most of the regions frequented by criminal cases involving environmental pollution, such as Zhejiang,[10] Hebei,[11] Shandong,[12] Jiangsu[13] and Henan,[14] have introduced normative documents to set forth explicit stipulations for the issues concerning the handling of such cases. In turn, the normative documents unify the understanding and are conducive to investigating, transferring and hearing relevant cases.

2.3 The Criminal Cases Involving Environmental Pollution were Intensively Governed by the Provisions of the Interpretation

All criminal cases involving environmental pollution heard and closed in Zhejiang Province in 2014, except particular cases, were governed by Paragraph 1–Paragraph 5, Article 1 of the Interpretation concerning conviction and sentencing.[15] Paragraph 3 from these five paragraphs was the most extensively applied, and the circumstance provided in Paragraph 3 is that "the illegally discharged pollutants which contain heavy metals and persistent organic pollutants and severely harm the environment and human health exceed three times the national pollutant discharge standard or the pollutant discharge standard specified by the people's governments of provinces, autonomous regions and municipalities directly under the Central Government according to legal authorization." In most of the environmental pollution cases heard and closed in Zhejiang Province, the heavy metals in wastewater

---

10 Zhejiang Provincial High People's Court, People's Procuratorate, Public Security Department and Environmental Protection Department jointly issued the *Circular Concerning the Printing and Distribution of the Minutes of the Meeting Regarding Several Issues Involving the Handling of Criminal Cases Relating to Environmental Pollution*. 2014-05-16.

11 Hebei Provincial Environmental Protection Department, High People's Court, People's Procuratorate and Public Security Department jointly issued the *Circular Concerning the Printing and Distribution of the Opinions on Further Strengthening Joint Law Enforcement against the Crimes of Environmental Pollution*. 2014-08-21.

12 Shandong Provincial People's Procuratorate, Public Security Department and Environmental Protection Department jointly issued the *Circular Concerning the Printing and Distribution of the Minutes of the Meeting Regarding the Symposium on the Province-wide Handling of Criminal Cases Involving Environmental Pollution* (Lu Jian Hui [2014] No. 4).

13 Jiangsu Provincial High People's Court and People's Procuratorate jointly issued the *Implementation Opinions on Several Issues Concerning Law-based Handling of Environmental Protection Cases*. 2013-11-01.

14 Henan High People's Court, People's Procuratorate, Public Security Department and Environmental Protection Department jointly issued *Several Opinions on Law-based Handling of Criminal Cases Involving Environmental Pollution (Trial)* (Yu Gao Fa (2014) No. 118).

15 Liang Jian and Ruan Tiejun, "Ascertaining of Other Severe Environmental Pollution Circumstances in the Crimes of Environmental Pollution," *People's Judicature*. 2014 (18).

discharged by the production operation personnel of electroplating processing enterprises exceeded three times the discharge standard, constituting the crime of environmental pollution.[16] For example, among 46 criminal cases involving environmental pollution heard by Ningbo Intermediate People's Court, Zhejiang Province from October 2013 to September 2014, 42 cases involved heavy metals exceeding three times the standard in sewage from small electroplating workshops and 1 case related to heavy metals exceeding the standard in sewage from a company.[17]

2.4 Those Criminalized Due to Environmental Pollution Fell Outside the Group of Enterprises above the Designated Scale

As shown by the cases across China and those in particular provinces with a large number of cases, the criminal cases involving environmental pollution mainly involved the owners and employees of small and micro enterprises, while the group of enterprises above the designated scale were seldom involved in these criminal cases. At the national level, after the Interpretation was implemented, 282 out of 1,104 people against whom judgments concerning the crimes of environmental pollution were entered were owners of private enterprises and individual workers, 496 were farmers and rural migrant workers, accounting for 25.54% and 44.93%, respectively (see Figure 8.4). In the criminal cases involving environmental pollution heard in Zhejiang, the defendants were basically owners engaged in individual operations and the unemployed. The causes for such a situation include the following: large enterprises and large companies allocated large sums of environmental protection funds, and were highly capable of making pollution treatment equipment available, while Zhejiang's economy was dominated by a private economy, Zhejiang's enterprises were mainly small and medium-sized, small and micro enterprises, more importantly, and the efforts to investigate the group of enterprises above the designated scale with environmental pollution behavior were insufficient to some extent.[18] Of course, with the implementation of the Interpretation, the authorities in different areas have pushed ahead with cracking down on illegal criminal activities involving environmental pollution,

---

16   Liang Jian and Ruan Tiejun, "Ascertaining of Other Severe Environmental Pollution Circumstances in the Crimes of Environmental Pollution," *People's Judicature*. 2014 (18).
17   See the *White Paper of Ningbo Intermediate People's Court Concerning Judgments of Criminal Cases Involving Disruption of the Protection of the Environment and Resources*.
18   See the speech delivered by Chen Guangduo, Presiding Judge of No.1 Criminal Tribunal, Zhejiang Provincial Higher People's Court, at the news briefing "Zhejiang's Courts Act on Treating Sewage, Preventing Flooding, Discharging Stagnant Water, Guaranteeing the Water Supply and Saving Water to Build a Beautiful Zhejiang and Create a Good Life According to Laws" in June 10, 2014.

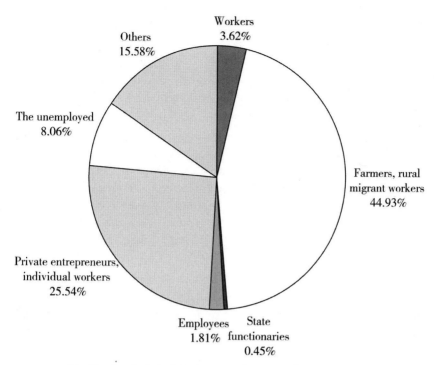

FIGURE 8.4   The identity of criminals in the criminal cases involving environmental pollution from July 2013 to October 2014

the cases in which the group of enterprises above the designated scale committed the crimes involving environmental pollution have started to appear and have gradually increased in different areas, especially in the area of crimes concerning hazardous wastes.

2.5   *Those Committing Hazardous Waste Crimes were Severely Punished*

In practice, it was very common that many enterprises knew that others did not obtain a business certificate or went beyond their business certificate, and provided others with hazardous wastes or entrusted others to collect, store, utilize, dispose of hazardous wastes, so as to reduce their own costs for disposing of hazardous wastes. After receiving hazardous wastes, others often directly dump those wastes into the soil and rivers due to a lack of the corresponding capability for disposal, thus severely polluting the environment. Given the amount of money paid for this service,[19] relevant departments were

---

19   The underlying cause for the hazardous waste crimes was that the production enterprises which generated hazardous wastes sought exorbitant profits. According to relevant information from Zhejiang, the normal cost for disposing of the pollutants from industrial

often aware of this fact, but let the actual result of severe environmental pollution go unchecked. However, before the Interpretation was implemented, not many enterprises were subject to criminal liability for the crime of major accidents involving environmental pollution—which was later changed to the crime of environmental pollution—because there were difficulties in ascertaining the joint criminal intention of enterprises. In response to this situation, Article 7 of the Interpretation specially provides that, where an actor knows that others have not obtained a business certificate or go beyond their business certificate, and provides others with hazardous wastes or entrusts others to collect, store, utilize, dispose of hazardous wastes so that the environment is severely polluted, that actor commits a joint crime of environmental pollution. According to this article, the authorities in different areas ferreted out and carefully investigated the hazardous waste crimes, focused on fighting crimes at the source, finding those behind the scene, and thus produced a positive effect after the Interpretation was implemented. The environmental pollution case of Zhejiang Huidelong Dyes & Chemical Co., Ltd.—the first environmental protection case in Zhejiang—is an example.[20] The series of environmental pollution cases of Tonglu Jinfanda Biochemical Co., Ltd. and

---

production, such as hazardous wastes, toxic substances and solid wastes, according to environmental protection requirements was 2,800 yuan–3,200 yuan/(t·trunk), but the cost was 60 yuan–120 yuan/(t·trunk) where relevant production enterprises entrusted others to illegally dispose of these pollutants; if these enterprises directly discharged these pollutants, the cost for disposing of these pollutants was near zero, so these enterprises would earn staggering profits in the case of illegally disposing of pollutants, they became the most important gainer from crimes regarding environmental pollution. See the speech delivered by Chen Guangduo, Presiding Judge of No.1 Criminal Tribunal, Zhejiang Provincial Higher People's Court, at the news briefing "Zhejiang's Courts Act on Treating Sewage, Preventing Flooding, Discharging Stagnant Water, Guaranteeing the Water Supply and Saving Water to Build a Beautiful Zhejiang and Create a Good Life According to Laws" in June 10, 2014.

20 Zhejiang Huidelong Dyes & Chemical Co., Ltd. illegally discharged, disposed of and dumped more than 23,000t of hazardous wastes via companies and individual businesses without the qualification for disposal of hazardous wastes. On June 30, 2014, the people's court of Shangyu District, Shaoxing City entered the first instance judgment that Huidelong was convicted of the crime of environmental pollution and was imposed with a fine of 20 million yuan, and 11 defendants were sentenced to criminal detention for 6 months and fixed-term imprisonment for 4.6 years. According to relevant information, this was so far the environmental pollution case with the highest amount of fines in Zhejiang Province. See Shaoxing, "Building an All-round Judicial Protection Chain for Water Environment," *People's Court Daily*. 2014-07-15, pg 3.

the environmental pollution case of Wynca Chemical[21] further indicate that Article 7 of the Interpretation has fully played its role in practice and the effect of its application is very obvious.

2.6 The Crime of Illegally Disposing of Imported Solid Wastes was Activated

During the period from the implementation of the *Criminal Law*, in 1997, to the implementation of the Interpretation, there were no cases involving the crime of illegally disposing of imported solid wastes or of the crime of importing solid wastes without authorization governed by Article 339 of the *Criminal Law* in practice. After the Interpretation was implemented, the criminal cases in which people were sentenced for the crime of illegally disposing of imported solid wastes started to occur. As mentioned above, 6 people were sentenced for the crime of illegally disposing of imported solid wastes.

2.7 The Decline in the Relative Number of the Criminal Cases Involving Dereliction of Duty in Environmental Regulation was Very Noticeable

In recent years, the cases involving the crime of a major accident involving environmental pollution (later changed to the environmental pollution crime) were almost as many as those involving the crime of dereliction of duty in environmental regulation, their ratio was nearly 1:1; in other words, once a case involving a major accident regarding environmental pollution was investigated and handled, a case involving dereliction of duty in environmental regulation was certainly investigated and handled, even only the criminal liability for dereliction of duty in environmental regulation was imposed, while no criminal liability for environmental pollution behavior was imposed. For example, from

---

21 The effective judgment documents have not yet been available. They are so far the two largest environmental pollution cases solved in Zhejiang Province. Jinfanda Biochemical and Wynca Chemical are leading domestic enterprises specializing in glyphosate production and key production enterprises in Zhejiang Province. According to the public security organ, since November, 2011, a subsidiary of Jinfanda Biochemical has transferred the hazardous wastes from agricultural chemical production to a company without the disposal qualification to illegally dump more than 38,000t of hazardous wastes in order to reduce production costs. Since May, 2012, Jiande No. 2 Chemical Factory of Wynca Chemical has delivered, through Jiande Hong'an Freight Transport Co., Ltd., more than 10,000t of phosphate mixture (a hazardous waste) to Rongsheng Chemical Co., Ltd. at Yuhang District, Hangzhou City, while Rongsheng Chemical directly discharged more than 7,500t of phosphate mixture to the Beijing-Hangzhou Grand Canal, severely polluting that body of water. See "Two Glyphosate Giants Illegally Dumped Several Tens of Thousands of Liquid Wastes," Zhejiang News in *Dushikuaibao*. 2013-12-17.

2008 to 2010, the ratio of the cases involving the crime of a major accident regarding environmental pollution to that involving dereliction of duty in environmental regulation was 11:13, 18:23 and 19:15, respectively. From 2011 to 2013, that ratio was changed to 24:11, 32:14 and 104:12, respectively. From July 2013 to October 2014, the people's courts heard and closed 23 criminal cases involving dereliction of duty in environmental regulation, effective judgments were entered against 26 people, the ratio of the cases involving the environmental pollution crimes to that involving dereliction of duty in environmental regulation was 743:23, and the ratio involving the number of people against whom effective judgments were entered was 1,104:26, showing a great difference between the two crimes. The reasons for the above situation can be broadly summarized as follows: First, the criminal cases involving major accidents regarding environmental pollution (later changed to environmental pollution) mostly occurred with the group of enterprises above the designated scale where major accidents involving environmental pollution occurred before the Interpretation was implemented, while the environmental pollution behavior of these enterprises were associated with dereliction of duty committed by environmental regulation personnel. After the Interpretation was implemented, those subject to criminal liability due to the environmental pollution crimes mostly fell outside the group of enterprises above the designated scale, while there were fewer associated with the environmental regulation personnel. Second, after the Interpretation was implemented, given that the situation had changed, the environmental regulation personnel realistically enhanced their responsibility awareness, performed the duty of environmental regulation, the circumstances where dereliction of duty in regulation occurred decreased.

3   Issues to be Urgently Reinforced in the Implementation of the Interpretation

The implementation of the Interpretation has produced a very obvious effect, so it should be fully recognized. However, as indicated by the implementation of the Interpretation for more than one year, some issues need to be addressed or further reinforced. Only in this way can the Interpretation play a more important role in fighting the crimes involving environmental pollution and safeguarding the ecological civilization.

3.1 Unify the Application of the Law in Fighting the Crimes Involving Environmental Pollution

The implementation of the Interpretation has also exposed some difficulties in the application of the law, such as handling of attempted illegal discharge, dumping and disposal of hazardous wastes, identification of hazardous wastes, identification of the cases where pollutant discharge exceeds three times the pollutant discharge standard, the scope of public and private property loss and the scope of heavy metals. In the practical handling of cases, relevant departments have different understandings of these issues, thus affecting the effect of cracking down on the crimes involving environmental pollution. Some local authorities have unified provisions at the regional level through normative documents, but these issues are widespread nationwide. They should be unified by the supreme judicial organ together with relevant departments.

3.2 Strictly Enforce Laws and Judicial Interpretations

After the Interpretation was issued, strictly enforcing it is the key and difficult point. In more than one year after the Interpretation was implemented, the cases in which the group of enterprises above the designated scale was subject to criminal liability for crimes involving environmental pollution occurred in small quantities, and no or few cases took place in only a few areas. The reasons were complicated, but undeniably, this was related to intensified investigation of illegal criminal activities involving environmental pollution. There is no privilege before the environment. It is urgently necessary to improve the fairness in the application of the *Criminal Law* during the implementation of the Interpretation. It is essential to treat various players, including the group of enterprises above the designated scale, equally without discrimination and strictly enforce laws to enhance the credibility of the judicial interpretation, further increasing the active role of the Interpretation in promoting the construction of an ecological civilization.

3.3 Build and Improve a Long-Term Mechanism for Preventing Environmental Pollution

Criminal penalty is imposed in order to avoid further criminal penalty. The real objective of the Interpretation is to give full play to the functions of criminal penalty in deterrence and education, reduce and prevent crimes involving environmental pollution. Therefore, actions should be taken to reinforce publicity concerning the Interpretation, further foster the atmosphere of public opinion that "environmental pollution will lead to imprisonment," and intensify the awareness of the law among the people so that the people from all walks of life, especially key environmental protection enterprises, really understand the provisions and value orientation of the Interpretation and effectively avoid criminal risks involving environmental pollution in their day-to-day operations.[22] Meanwhile, it is also necessary to avoid movement-type law enforcement and undue expansion of the scope so as to ensure that the criminal policy of combining punishment with leniency is practically implemented in the environmental criminal judicial field.

---

22   Many operators only concentrate on making more money without paying attention to other things. After the Interpretation was implemented, some environmental polluters did not realize that they were suspected of committing crimes. For example, when the law enforcement personnel at the minor crime division of the environmental protection bureau conducted investigation, Mr. Kang, the legal representative of Wenzhou Xingshi Optical Co., Ltd. in Wenzhou lightly said that based on experience, "I thought this time I needed to pay a penalty of about 40,000 yuan, and I had prepared it". See "All-out Efforts Were Made in Law Enforcement, No Law-breakers Can Evade Legal Sanctions," *China Environment News*. 2014-07-23, pg 1.

CHAPTER 9

Environmental Information Disclosure Made Breakthroughs in 2013–2014

*Wu Qi**

Abstract

A series of laws and regulations regarding environmental information disclosure were enacted and implemented in 2013–2014, greatly perfecting the institutional framework and making breakthroughs in many areas. In 2014, the Supreme People's Court published for the first time *Ten Cases Regarding Government Information Disclosure*, which addressed some debatable issues and clarified crucial details, and therefore carried great significance in striving for environmental information disclosure. However, the new system must rely on the effective law enforcement and careful information processing to serve its irreplaceable role in environmental protection.

Keywords

environmental information disclosure – breakthroughs in priority areas – judicial protection and public supervision

The 2014 *Amendment to the Environmental Protection Law* (the Amended law is hereinafter referred to as EPL) offered new possibilities in tackling environmental problems shrouded in smog. While the newly amended EPL has an independent chapter on environmental information disclosure and public participation, confirming their crucial roles, specific issues remain unsettled as to effective law enforcement and better utilization of disclosed information.

Enforced on May 1, 2008, *Measures on Open Environmental Information (for Trial Implementation)* (hereinafter referred to as the *Measures*) is a milestone in the progress of China's environmental regulation. The following seven years have witnessed the step-by-step development of information disclosure system

---

* Wu Qi is the staff attorney of the Beijing Office of Natural Resources Defense Council (NRDC). This article was written for the project of China Environmental Law conducted by NRDC.

from scratch. To establish a baseline of performance and evaluate its development, the Institute of Public and Environmental Affairs (IPE) has cooperated with the Natural Resources Defense Council (NRDC) to launch the Pollution Information Transparency Index (PITI). The annual PITI evaluation of 113 cities of the *Measures* enforcement and disclosure of pollution source monitoring data indicates that the preliminary system of environmental information disclosure has already been established in the past few years. The evaluation also indicates, that while overall performance has gradually improved, different cities vary greatly in the scope of information disclosure. The disclosure of some crucial information, emission data and environmental impact assessment for instance, is inadequate due to a lack of legal requirement or operational standard. Making breakthroughs in these priority areas; therefore, would greatly propel the disclosure of environmental information.

Since the end of 2012, continuous smoggy days have triggered the accelerated disclosure of environmental information. The Ministry of Environmental Protection[1] (is hereinafter referred to as MEP) has launched a series of regulations and policies to expand the scope of environmental information disclosure and strengthen it from all angles, perfecting the institutional requirement. A relatively complete legal framework has been gradually formed, and substantial breakthroughs have been made in such priority areas as pollution source monitoring, emission data, and environmental impact assessment.

1 Breakthroughs in Priority Areas

1.1 *The System of Governmental Disclosure of Pollution Source Monitoring was Strengthened*

The *Measures* provides the basic framework of the disclosure system of environmental information in China. As the first law specializing on disclosure system, the *Measures* leaves room for improvement. Some regulations and principles are practiced inconsistently in reality.

On the basis of the Measures, The Ministry of Environmental Protection issued *Announcement of Strengthening of Environmental Pollution Sources Monitoring Information Disclosure Work* (*the Announcement*) in July 2013. With the specification of such details as providers, content, methods, platforms of information, etc., the system of pollution information disclosure is regulated. The categories of pollution information specified in the *Announcement*

---

1 The Ministry of Ecology and Environment superseded the Ministry of Environmental Protection (MEP) in 2018.

covered basic information of intensive monitoring of pollution source, control of total volume, prevention of pollution, collection of pollutant discharge fees, supervision and law enforcement, administrative penalty, environmental contingency planning, evaluation of enterprise environmental behaviors, etc. The *Announcement* also requires all levels of environmental protection authorities to fully release information of pollution sources monitoring on the platform of government websites with a user-friendly interface and convenient search function. Furthermore, disclosure of pollution source monitoring should be released voluntarily in a specialized column of the website. *The Annex to the Announcement—The List of Pollution Sources Information Disclosure* (1st Batch)—provides technical specifications such as content, time limits and provider of information, and therefore serves as a guideline on how to enforce the system.

In April 2014, *the Key Points in the Government Information Disclosure Work* issued by State Council again lists environmental information disclosure as a priority area of government work for three successive years. Environmental information also ranked 1st on the list of categories of public monitoring information disclosure. According to the document, priority issues include air and water quality, environmental impact evaluation of construction projects, pollution sources monitoring, pollution sources monitoring information provided by state-controlled key enterprises and pollution reduction.

1.2 Breakthroughs in Disclosing Pollutants Emission Data

The scope of corporate mandatory disclosure of environmental information set in the *Measures, Cleaner Production Promotion Law*, and other supporting regulations is still relatively limited. In terms of emission data, enterprises are obliged to undertake mandatory responsibility of disclosing pollutants emission data only when their emission of pollutants exceeds the standard, or their total volume of emission exceeds the quota, or major and extraordinarily big environmental contamination accidents occurs. In contrast, two regulations that were issued recently make substantial breakthroughs in terms of emission data disclosure.

Measures for the Environmental Management Registration of Hazardous Chemicals (for Trial Implementation)[2] issued by the Ministry of Environmental Protection in 2012 made it an obligation for the first time for enterprises to disclose information on release and transfer of priority hazardous chemicals and characteristic chemical pollutants, thus forming the basic framework of a Chinese version of PRTR (Pollutant Release and Transfer Register). Its

---

2 Translated in pkulaw.cn, http://www.lawinfochina.com/display.aspx?lib=law&id=12234.

Article 22 stipulates that enterprises producing and using hazardous chemicals should issue an annual report to the public on environmental management of hazardous chemicals each January. This report should include the types of hazardous chemicals, hazardous properties, release and accident information of related pollutants. pollution preventive measures. Enterprises producing or using hazardous chemicals subject to priority environmental management are also required to publish information of release and transfer of the chemicals and monitoring results. According to this rule enforced on March 1, 2013, the first batch of release data disclosure should have been published by January 2014, but it was postponed till April 2014, when the *Inventory of Hazardous Chemicals* subject to priority environmental management was issued by the MEP in April 2014, laying the foundation for data disclosure.

In July 2013, *the Measures for the Self-Monitoring and Information Disclosure by the Enterprises subject to Intensive Monitoring and Control of the State (for Trial Implementation)* and *the Measures for the Pollution Sources Supervisory Monitoring and Information Disclosure by the Enterprises subject to Intensive Monitoring and Control of the State (for Trial Implementation)*[3] were issued. Respectively, these two regulations made it explicit that enterprises have the obligations to self-monitoring and information disclosure and environmental protection departments are responsible for supervisory monitoring of these enterprises. These two *Measures* also initiated a new phase of online monitoring and real-time information release. More than twenty provinces and independent municipalities have set up platforms for information disclosure by the enterprises subject to intensive state monitoring and control since 2013. Such provinces as Shandong, Zhejiang and Jiangxi have set up effective systems and platforms of online monitoring and real-time information release. Real-time release of online monitoring data, in particular, has strategic significance, as it not only discloses stealthy pollutants discharge of enterprises by exposing it to public monitoring, but also helps to identify pollution sources and promotes inter-regional joint prevention and control.

1.3 Full Disclosure of Environmental Impact Assessment (EIA)

As a necessary prerequisite to public participation, disclosure of EIA has always been the focus of various environmental disputes. Additionally, the disclosure of information, especially the environmental protection measures promised by enterprises in EIA reports, is the basis of continuous public scrutiny.

---

3 Translated in pkulaw.cn, http://en.pkulaw.cn/display.aspx?id=523a5bfb4d496e22bdfb&lib=law.

In November 2013, the environmental ministry promulgated *Guide for Open Government Information on Construction Project Environmental Impact Assessment (for Trial Implementation)* (the *Guide*), specifying the disclosure of EIA document of construction project subject to government approval, acceptance and administration. Furthermore, the *Guide* made breakthroughs in full disclosure of EIA document. According to the *Guide*, before developers of construction projects submit EIA reports and statements to the environmental protection authority, they are required to voluntarily prepare a complete version including an appendix of the deleted contents concerning state and commercial secrets and justification for the deletion. The environmental protection authority is required to audit the appendix upon acceptance and disclose the complete, unabridged reports and statements of EIA in accordance with the *Guide*. Full disclosure of EIA has made significant progress in some regions, providing an essential basis for strengthening public participation in the process and aftermath of EIA. Nevertheless, full disclosure of EIA is only the first step; it requires both substantial public participation and mutual communication and negotiation to resolve related NIMBY (Not in My Back Yard) disputes.

1.4 *The Revised Environmental Protection Law Guarantees New Progress*
On April 26, 2014, the revised *Environmental Protection Law* was adopted twenty-five years after its first promulgation. The newly amended EPL has an independent chapter on environmental information disclosure and public participation, with specific articles guaranteeing the new development in the above-mentioned sections. Article 55,[4] in particular, stipulates that key pollutant-discharging units shall truthfully disclose the names of their major pollutants, the ways of emission, the emission concentration and total volume, the standard-exceeding emission status. To aid the implementation of the amended EPL, MEP promulgated further administrative measures—*Measures for the Disclosure of Environmental Information by Enterprises and Public Institutions (For Trial Implementation)*, specifying the scope of disclosure of key pollutant-discharging units. This order was issued in December 2014 and came into force in January 1, 2015 to aid the implementation of the newly revised EPL. Article 56[5] of the amended EPA also stipulates that the competent department responsible for the examination and approval of EIA documents shall make public the full text of environmental impact reports of the construction

---

4 Translated by China Environmental Governance Programme, https://www.chinadialogue.net/Environmental-Protection-Law-2014-eversion.pdf.
5 Ibid.

project with exception of commercial secrets and confidential circumstances as specified by the State.

Also noteworthy is the addition of individual liability in the case of violating information, particularly tempering or forging the monitoring data. Article 86[6] specifies that persons directly in charge and other personnel are subject to direct liabilities. The nine violation acts specified include tampering or forging monitoring data or instigating others to do so, failing to disclose environmental information that should be disclosed in accordance with the law. In addition, Article 63 stipulates that those who commit tampering or forgery of monitoring data among other acts which violate laws shall be imposed a detention.

2   The Legal Guarantee

Laws alone cannot carry themselves into practice. Controversial issues remain where environmental information disclosure is enforced. *The Regulation of the People's Republic of China on the Disclosure of Government Information* has been carried out since May 1, 2008, and the past seven years have witnessed an increasing number of lawsuits concerning environmental information disclosure. In September 2014, the Supreme People's Court for the first time published *Ten Cases Regarding Government Information Disclosure*, addressing major debatable issues such as in-process information, inside information, definition of trade secrets and enterprise environmental information acquired by government. Most of these cases are closely related to environmental information disclosure. Some major debatable issues[7] involved in applying for open environmental information are listed as follows:

2.1   *Whether the Enterprise Information Acquired by Administrative Bodies Should Be Counted as Government Information.*

Since the *Measures* was carried out, departments of environmental protection have often rejected applications for information disclosure on the ground that environmental impact information is possessed by enterprises. The Supreme People's Court addressed this controversial issue in the case of *Yu Suizhu v. Sanya Municipal Bureau of Land and Environmental Resources, Hainan Province*. In this case, Sanya Municipal Bureau of Land and Environmental Resources held that the report form for the EIA under the application of the

---

6   Ibid.
7   Zhou Bin and Liu Ziyang, "The Supreme People's Court Published *Ten Typical Cases Regarding Government Information Disclosure*," Legaldaily.com, September 12, 2014, http://www.legal daily.com.cn/xwzx/content/2014-09/12/content_5760759.htm.

plaintiff for disclosure was a document of the enterprise other than government information and thus should not be disclosed. The Court defined that "externally-acquired information is also government information."

> The government information included not only information that was prepared by an administrative organ, but also information that is acquired by the administrative organ from citizens, legal persons, or other organizations. Therefore, the environmental information of the enterprise that was acquired by the administrative organ in the process of performing duties was also government information.[8]

When the Ministry of Environmental Protection issued *Guide for Open Government Information on Construction Project Environmental Impact Assessment (for Trial Implementation)* and the new *Amendment to EPL*, the obligation of environmental protection authorities to disclose complete report of EIA is specified. More significantly, by the ruling of *Yu Suizhu v. Sanya Municipal Bureau of Land and Environmental Resources*, the Supreme People's Court defined the nature of enterprise environmental information acquired by the administrative organ in the process of performing duties as government information—the enterprise environmental information is not only restricted to EIA document, but also includes other information related to pollutants release and environmental performance—therefore served as an important guide to future act decisions.

2.2 Whether the Information Applied for Disclosure Should Be Counted as Trade Secrets

The trade secret is another debatable issue in the practice of applying for environmental information disclosure. The current laws and regulations do not give a clear definition of what should be counted as trade secrets; therefore, it is difficult for the applicants, environmental protection authorities and judicial courts to determine the nature of environmental information accordingly. As for this issue, the case of *Wang Zongli v. Real Estate Administrative Bureau of Heping District, Tianjin Municipality* can be referred to in the future. This case did not involve environmental information, but the controversy was focused on the disclosure of government information that involved trade secret. The Supreme People's Court held that in the practice of government information disclosure, the administrative organs constantly reject the application on the ground that the government information under application for disclosure involves trade secrets, but such administrative organs sometimes

---

8 Translated in pkulaw.cn, http://lawinfochina.com/display.aspx?id=18035&lib=law.

abuse the power. Furthermore, it pointed out that the concept of trade secret has strict connotation. "In accordance with the Anti-Unfair Competition Law, trade secret means the practical technical information and business information which is unknown by the public, which may create business interests or profit for its legal owners, and also is maintained secrecy by its legal owners. An administrative organ should examine an application of information disclosure according to this standard, ... During the legality review, a people's court should enter a judgment on whether it is a trade secret according to evidence presented by an administrative organ."[9] According to the ruling, the burden of proving whether the information should not be disclosed on the ground of trade secrets is now required of the administrative organ.[10]

2.3 Whether Procedural Information and Historical Information Should Be Disclosed

In the practice of environmental information disclosure, it is difficult to decide whether certain types of information are subject to public disclosure, as there is no clear definition. For instance, in-process information, inside information of administrative organs, archived materials and historical information prior to the implementation of *Measures on Open Environmental Information (for Trial Implementation)* all belong to this category. The Supreme People's Court also demonstrated an approach to dealing with such information with the ruling of the following case.

In the case of *Yao Xinjin and Liu Tianshui v. Yongtai Municipal Bureau of Land, Fujian Province*, the Court ruled that "In-process information generally refers to information on study, deliberation, request for instructions and report that is internal within an administrative organ or is formed among administrative organs. The uniform disclosure or earlier disclosure of such information may impair the integrity of the decision-making process and the effective handling of administrative affairs. However, the in-process information should not be an absolute disclosure exception. After the decision-making was completed or a decision was made, previous information under investigation, deliberation, and handling was no longer in-process information. If the disclosure demands were greater than the non-disclosure demands, such information should be disclosed."[11]

---

9   Translated in pkulaw.cn, http://lawinfochina.com/display.aspx?id=18035&lib=law.
10 Zhou Bin and Liu Ziyang, "The Supreme People's Court Published Ten Typical Cases Regarding Government Information Disclosure," Legaldaily.com, September 12, 2014, http://www.legaldaily.com.cn/xwzx/content/2014-09/12/content_5760759.htm.
11 Zhou Bin and Liu Ziyang, "The Supreme People's Court Published Ten Typical Cases Regarding Government Information Disclosure," Legaldaily.com, September 12, 2014, http://www.legaldaily.com.cn/xwzx/content/2014-09/12/content_5760759_2.htm.

The main issue of the case of *Qian Qunwei v. People's Government of Zhangqi Township, Cixi City, Zhejiang Province* is the disclosure of historical information. By definition, historical information refers to the government information that has been formed before the implementation of *Regulations on Open Government Information of the People's Republic of China*. The determination in the judgment of this case that "the contention of the defendant that the government information before the implementation of the Regulation could not be disclosed lacked legal basis" conformed to the legislative intention. The principle of "non-retroactivity of law" means that the provisions of legal documents only apply to events and acts after such legal documents come into force and they do not apply to events and acts before such legal documents come into force. As far as this case was concerned, the so-called event and act referred to the application of the plaintiff for the disclosure of government information in accordance with the Regulation and the reply on the application given by the administrative organ. The judgment of this case that "the Regulation has already been implemented when the plaintiff applied for the disclosure of government information" was a correct understanding of the principle of "non-retroactivity of law."[12] The principle clarified in this case proves to be significant for the future practice of environmental information disclosure, especially environmental public litigation, as it often fails to request relevant information disclosure and therefore provide sufficient evidence.

2.4 *The Applicant's Description of Requested Information Should Be as Detailed as Possible*

Whether the applicant's description and definition of requested information match those possessed by administrative organs remains an issue in the system of environmental information disclosure. On one hand, if the applicant's description of requested information is too general, environmental protection authorities will find it difficult to meet their needs as they have to make extra effort to investigate, to collect, or to sort out information. On the other hand, applicants always have difficulty defining or describing the requested information because they don't have archive reference numbers or other relevant clues.

Translated in pkulaw.cn, http://en.pkulaw.cn/display.aspx?id=e53b14f273235051bdfb&lib=law.

12 Zhou Bin and Liu Ziyang, "The Supreme People's Court Published Ten Typical Cases Regarding Government Information Disclosure," Legaldaily.com, September 12, 2014, http://www.legaldaily.com.cn/xwzx/content/2014-09/12/content_5760759._4.htm. *Translated in* pkulaw.cn, http://en.pkulaw.cn/display.aspx?id=e53b14f273235051bdfb&lib=law.

The ruling of *Zhang Liang v. Shanghai Bureau of Planning and Land Resources* established two principles relevant to such cases. First, when submitting an application for the disclosure of government information, an applicant should describe such government information as detailed as possible so that it will facilitate the administrative organ to conduct the retrieval. Second, the administration organ shall not provide government information that does not exist.

> Where an administrative organ refuses to provide government information on the ground that such information does not exist, the administrative organ should prove that it has performed reasonable retrieval duties; and as for the description of an applicant on the information under his or her application, the applicant should not be demanded to specify the standard title of such government information or even the specific numbers of document issuance. If the administrative organ conducted retrieval only with the keywords based on the plaintiff's description and simply replied that the government information did not exist, the administrative organ did not perform its retrieval duties.[13]

2.5 *The Relationship between Voluntary Disclosure and Disclosure upon Request*

According to *Measures for the Disclosure of Environmental Information (for Trial Implementation)* and other related laws and regulations issued recently, environmental protection authorities are obliged to voluntarily disclose a fairly wide scope of environmental information. However, the question arises when it comes to the relationship between voluntary disclosure and disclosure upon request—for government information that has been voluntarily disclosed, should an administrative organ be allowed to disclose it repeatedly? The case of *If Love Matrimonial Service Co., Ltd. v. Ministry of Civil Affairs of the People's Republic of China* clarified the relationship between voluntary disclosure and disclosure upon request, though it did not involve environmental information. According to the case analysis of the Supreme People's Court, "the methods for the disclosure of government information include voluntary disclosure and disclosure upon application, and these two methods are complementary. For government information that has been voluntarily disclosed,

---

13 Zhou Bin and Liu Ziyang, "The Supreme People's Court Published Ten Typical Cases Regarding Government Information Disclosure," Legaldaily.com, September 12, 2014, http://www.legaldaily.com.cn/xwzx/content/2014-09/12/content_5760759._4.htm. Translated in pkulaw.cn, http://en.pkulaw.cn/display.aspx?id=e53b14f273235051bdfb&lib=law.

an administrative organ may be allowed not to disclose it repeatedly; however, the administrative organ should notify the applicant of the method and way for access to such government information."[14]

3 An Effectively Functioning Information Disclosure System Requires Combined Efforts

The years of 2013 and 2014 saw important breakthroughs in the various aspects of establishing the institutional framework of environmental information disclosure system. The progress of legislation and legal guarantee, however, only provided the basis for environmental information disclosure. The new system must rely on the effective law enforcement and utilization of accessed information to truly make a difference. For instance, the newly issued *Regulation on Public Participation in Environmental Protection of Hebei Province* has a separate article stating that environmental protection organizations are encouraged to monitor the information disclosure and collect disclosed information in accordance with the law. In terms of social media, the Blue Sky mobile app[15] developed by the Institute of Public & Environmental Affairs (IPE) is great for the public to monitor and use environmental information. The users can consult the app to access the air quality data of their city and check the real-time emission monitoring data from key pollution sources, including the level of pollutants concentration, the standard limit, the exceeding rate and emission volume. In this way, they play a role of monitoring the disclosure of real-time data and scrutinize emissions from "major polluting companies." In Shandong and some other provinces, the local environmental protections organizations also began to interact with the polluting companies and supervisory departments by utilizing the disclosed real-time monitoring data, pressing the enforcement of environmental laws and the correction of unlawful acts.

With the implementation of 2014 *Amendment to the Environmental Protection Law*, we expect that environmental information disclosure will be normalized, the disclosed information will be used effectively, and the system will play a powerful role in curbing environmental and ecological damage.

---

14 Zhou Bin and Liu Ziyang, "The Supreme People's Court Published Ten Typical Cases Regarding Government Information Disclosure," Legaldaily.com, September 12, 2014, http://www.legaldaily.com.cn/xwzx/content/2014-09/12/content_5760759._5.htm. Translated *in* pkulaw.cn, http://en.pkulaw.cn/display.aspx?id=e53b14f273235051bdfb&lib=law.

15 The app was originally named "Pollution Map" when it was launched in 2006 and renamed "Blue sky Map" in 2015.

CHAPTER 10

China's New Urbanization Plan and Sustainable Consumption

*Chen Hongjuan and Chen Boping**

Abstract

This paper analyses the status quo, characteristics and trends of China's sustainable consumption in the context of *China Urbanization Plan 2014–2020*. The paper emphasizes the importance of sustainable consumption in China through comparison of ecological footprint and biocapacity, analyses the advocacy activities for reducing the ecological footprint, and then assesses the accelerated formula to promote sustainable consumption in the midst of the new wave of the national urbanization.

Keywords

urbanization – sustainable consumption – sustainable development – eco-cities

China Urbanization Plan 2014–2020 ("the new urbanization scheme" or *"the Plan"* or "the Scheme" hereinafter) was promulgated in early 2014, marking the issue of urbanization having been upgraded to the state strategy level. Aiming to boost domestic demand and to stimulate overall growth, the new urbanization scheme has also highlighted the importance of ecological civilization.

---

* Nancy Chen, a free-lance author, a senior environmental activist, and founder of ClimaXmi, an advisory firm based in the Netherlands. Ms. Chen started her environmental protection endeavor right after she earned her double Master's Degrees in Law and Humanitarian Assistance. She is specialized in climate change policies, energy related laws and she has been active in international researches and global project management of sustainable development.
  Boping Chen, China Director of the World Future Council (WFC), dedicated to WFC's Regenerative Cities Program in China. Previously, Ms. Chen worked for the WWF China, directing China for a Global Shift Initiative, leading general policy researches and advocacy for environmental dialogues on China's domestic development policy as well as sustainability policies for China's overseas investments.

The reaction to *the Plan* varies. On the one hand, the press as well as provincial and municipal governments were busy evaluating the impact of *the Plan* on land, fiscal revenue and household registration system; on the other hand, independent think tanks and not-for-profit organizations were seizing the opportunities to promote sustainable urbanization, namely environmental protection, energy conservation, and improvement of people's livelihood. Among other efforts, were World Wide Fund for Nature (WWF)'s annual *China Ecological Footprint Report* and its promotion of the Sustainable Consumption Week, which have provided the framework guideline and best practices of sustainable consumption from the macro-economic and final consumer perspectives.

1   Sustainable Consumption and the New Type of National Urbanization

1.1   *The Concepts and Interrelations of These Two Terms*

The definition proposed by the 1994 Oslo Symposium[1] on Sustainable Consumption defines it as "the use of services and related products which respond to basic needs and bring a better quality of life while minimizing the use of natural resources and toxic materials as well as emissions of waste and pollutants over the life cycle of the service or product so as not to jeopardize the needs of future generations."

Afterwards, the idea of global management of sustainable consumption started to appear in the reports and projects of the international organizations. The international academic community also began to invest more resources in related researches at the turn of the century. However, the focus of these efforts was more to suggest change of consumption behavior and policy frameworks at high level, than to create a holistic theoretical foundation or to systematically study the quality, structures and behavior of the sustainable consumption by country or by region.

The academia seems behind in terms of the researches of sustainable consumption. Although some of the academics in China are shifting attention from green consumption to sustainable consumption, they have not achieved breakthrough,[2] until the introduction of the concept of ecological footprint,

---

1   http://www.iisd.ca/consume/oslo004.html.
2   http://baike.baidu.com/view/1817544.htm.

which not only makes it possible to quantify the macro ecological footprint, but also to precisely examine how an individual's ecological footprint interacts with biocapacity.

The promulgation of the new urbanization scheme marks that new types of urbanization are attached as high importance as to strategic development of the nation. *The Plan* states that the future urbanization shall fit in the new trends of modern city development, such as promotion of greener, smarter cities, emphasis on cultural and historic heritage, and enhancement of the intrinsic qualities of the cities. In the section of "Accelerating green city development," *the Plan* calls for integration of ecological civilization into urbanization, and encourages green production, green lifestyle and green consumption behavior.[3]

Urbanization usually brings higher household income, which in turn leads to higher consumption, change of consumption behavior and increase of the ecological footprint. China's high demand for energy, raw materials and resources from the international market testifies that the nation's ecologic resources and energy are in short supply.

It is apparent that China's central government has come to realize that the nation's industrialization and urbanization has been at the sacrifice of natural environment. The land-based fiscal tools are simply not sustainable. Therefore, the new urbanization scheme stresses that the urbanization shall meet the requirements for ecological civilization, low carbon emission and energy conservation, that green production and green consumption shall set the trend for urbanization, and that China will prioritize ecological civilization in the course of urbanization, rather than duplicate developed countries' way of urbanization via high energy consumption and high emission. In this way, China will make great contributions to the global ecological safety and collect valuable experience in sustainable urbanization for developing countries.[4]

1.2 The Status Quo, Features and Trends of China's Sustainable Consumption

In 2007, as Chinese scholar Si Jinluan pointed out in the article "Sustainable Consumption: Theoretical Innovation and Policy Choices," while in US, Germany, Italy and the Netherlands, there are respectively 77%, 82%, 94% and 67% of the consumers who will consider ecological factors in their purchases,

---

3 http://news.xinhuanet.com/house/wuxi/2014-03-17/c_119795674.htm.
4 http://www.gov.cn/zhuanti/xxczh/.

in China, less than 20% of the people think about sustainable consumption and in addition, the sustainable consumer products market is flooded with shoddy and confiscated goods.[5]

In 2012, Global Sherpa conducted a survey on the consumers from 17 developed and developing countries and published "BRIC Countries Top Survey of Green Consumers." In this survey, guilt was included as one of the indicators. The Chinese consumers expressed second highest sense of guilt over the environmental implications of their behavior, only next to India.[6]

In 2013, China Consumer Association, Chinese Academy of Social Sciences, and L'Oreal jointly published *China Sustainable Consumption Research Programme Report 2012*, which surveyed in six major Chinese cities for the status quo of sustainable consumption, as well as consumers' awareness of and action towards the sustainable consumption.

Although the articles and surveys mentioned above are yet to depict a full picture of China's sustainable consumption, we can draw such conclusions as follows:

1. China is fairly far behind some developed countries in sustainable consumption;
2. Chinese urban population are becoming more rational and environmentally conscious in recent years than in 2007 when it comes to sustainable consumption;
3. Sustainable consumption has advanced considerably in some areas. For instance, in beauty care sector, the rate of sustainable consumption increased to 70–80% by 2013, from 20% in 2007 (assuming that the statistics are accurate and comparable);
4. China's service-oriented consumption lags behind many other countries.

Therefore, we have the ground to believe that under the new urbanization scheme in China, green consumption sets out to become new trend and new normal. Nevertheless, it requires the collaborative and consistent efforts from the governments of various levels, the NGOs from home and abroad, the business organizations as well as consumers in order to build a friendly circle for sustainable consumption.

---

5 http://qkzz.net/article/cb5f8697-2efb-4371-a4e5-ac7fed56e80a.htm.
6 Jason, "BRIC Countries Top Survey of Green Consumers", August 8, 2012, Global Sherpa, http://www.globalsherpa.org/green-consumer-research-sustainable-consumption.

2 Importance of Sustainable Consumption in China

There are three different ways to look at the importance of the sustainable consumption in China.

2.1 *The Status Quo of Biological Footprint and Biocapacity*

The WWF's *China Ecological Footprint Report 2014* ("the WWF Report" hereinafter) pointed out, "With rapid economic growth and urbanization, China's ecological system and natural resources are faced with increasing challenges. In particular, changes of the consumption behavior and the way we utilize natural resources, will have a significant impact on China's future."[7]

Since 1970s, the earth has entered ecological overshooting stage, which means every year humanity's resource consumption exceeds Earth's capacity to regenerate those resources. China is no exception, with its ecological footprint measured either by total or by per capita consumption along with the nation's economic boom. In 2008, China's ecological footprint per capita was 2.1 global hectare (gha), 2.4 times as much as its own biocapacity (0.87 gha).

2.2 *Urbanization and Ecological Footprint*

Urbanization serves as a key driver of China's ecological overshooting. *The WWF Report* has found out that China's urbanization rate jumped from 1980's 26% to 2011's 51.3% while urban population's consumption accounted for 80% of the nation's total consumption volume in 2011 compared to 40% three decades ago. *China Statistical Yearbook* indicates that the Chinese urban population's consumption power is three times equal to that of the rural population. As a result, rapid urbanization will only lead to more pressure on ecology and resources, and will become a crucial component of the consumption related ecological footprint.

The WWF Report also pointed out that the urban residential housing area has grown by 50% in the past decade and as a result, the housing related ecological footprint was accelerated. Meanwhile, the rapid increase of the cars has contributed to the rise of the transportation related ecological footprint. China's service sector, which is behind the high speed of the industrialization, will need favorable policies to stimulate sustainable consumption. All in all, accelerated urbanization will play a key role in driving the ecological footprint, of which is carbon footprint, in particular.

---

[7] http://www.wwfchina.org/content/press/publication/2014/CN2014footprint.pdf (Chinese version).

2.3 Significance of Changing Unsustainable Consumption Behavior Pattern

In accordance with the findings from the task force of "the Sustainable Consumption and Green Development" conducted by the China Council for International Cooperation on Environment and Development (CCICED), in some major Chinese cities, the natural resource consumption per capita has been growing rapidly, trending towards their counterparts in the industrialized countries in an accelerated pace.

This upward trending is reflected in the mounting ecological footprint per capita. China has felt enormous pressure in the United Nations' climate change negotiations in 2012, due to its world's highest total greenhouse emission and per capita emission which is almost as high as that in the developed countries. Therefore, China is determined to promote sustainable urban development and energy consumption, not only to ease the pressure from the international community, but also to better tackle the challenges of the climate change and energy crisis.

3 China's Policies, Laws, Regulations and Practices Pertaining to Sustainable Consumption

3.1 China's Policies, Laws and Regulations Pertaining to Sustainable Consumption

Following Rio De Janeiro's Agenda 21[8] since 1992, China's State Council approved the China Agenda for the 21st Century in 1994, providing high level strategy, plan and solutions for sustainable development.

In the new millennium, the Chinese governments at various levels continued to issue policies and guidelines regarding taxation, market mechanism, as well as industry adjustments and revitalization, which are favorable to sustainable consumption. Among others, China has promulgated over 70 green consumption policies concerning 11 major product categories. Unfortunately, these policies took little effect as they miss a holistic strategic solution for sustainable consumption, noting that the China Agenda for the 21st Century, due to its historical limitations, has put family planning and urban economic development as priorities.

Not only has China had put into place very few specialized laws or regulations regarding sustainable consumption, nor has the nation included any specific paragraphs promoting sustainable or green consumption in *the Environmental*

---

8   http://www.un.org/chinese/events/wssd/chap4.htm.

Protection Law, Water Law, Mineral Resources Law or *the Renewable Energy Law*. As a result, it is hard to find clear guidance on the objectives, definitions, scopes or responsible parties to promote sustainable consumption.

3.2   *Sustainable Consumption Practices for Reducing Ecological Footprint*

Consumption and lifestyle reflecting ecological civilization aims to protect natural ecological environment and maintain the balance of biocapacity. Therefore "sustainable" is the key to this consumption behavior while meeting human being's basic sustainment and development.

In the 21st century, China's practices in sustainable consumption have mostly been initiated by international and domestic environmental groups, consumers and business organizations, who seek to reduce the ecological footprint through industry and product upgrading.

For urban population, their ecological footprint is created by consumption around food, clothing, housing, transport and services. Urban consumers have put great efforts in exploring sustainable consumption in these five areas. This article will summarize consumer practices in food, clothing, housing and transport.

3.2.1   Clothing

China's sustainable consumption in clothing has made remarkable progress thanks to the efforts of environmental and animal protection organizations. For instance, Green Peace has launched "the Zero Discharge of Hazardous Chemical Programme." Twenty fashion brands were committed to meeting the goal of zero discharge of hazardous chemicals by January 2020.[9]

In addition, the consumers' awareness of green fashion has evolved from rejection of wearing animal fur, to appreciating natural materials such as cotton and hemp, then to rising interest in recyclable materials.

The Chinese urban consumers are increasingly aware that they need to know "who" "in what way" are making clothing for them, and how they can continue to use and reuse the materials and resources. Plenty of clothing recycling and alteration programs are unfolding in urban areas.

It is noteworthy that H&M, a Swedish fashion brand started a "Recycle Your Clothes" initiative worldwide to close the loop for fashion. In August 2013, the program was fully launched in all H&M stores in China. People can drop

---

9   The brands include Nike, Adidas, Puma, H&M, M&S, C&A, Li-Ning, Zara, Mango, Esprit, Levi's, Uniqlo, Benetton, Victoria's Secret, G-Star Raw, Valentino, Coop, Canepa, Burberry and Primark.

off their textiles, no matter brand or condition, in any of H&M stores. That year, H&M collected 3,047 metric tons of garments, and sort them into three categories:
- Rewear—clothing that can be worn again will be sold as second-hand clothes.
- Reuse—old clothes and textiles will be turned into other products, such as cleaning cloths.
- Recycle—everything else is turned into textile fibers, and used for things like insulation.

3.2.2 Food

In recent years, increasing importance was attached to the environmental implications of food production and consumption. In order to reduce such implications, the Food and Agriculture Organization of the United Nations (FAO) started to publish nutrition and food guidelines to promote sustainable dietary and biological diversities in China since 2010.[10]

In 2014, the WWF launched the Green Week in over 100 retail stores in Beijing, Shanghai, Guangzhou, Shenzhen, Hangzhou, Suzhou, Ningbo and Dalian, to promote the concept of the green consumption to consumers. The highlight of the Green Week was the Maritime Stewardess Council (MSC)'s themed program "Selecting Sustainable Seafood, Protecting Blue Marine Resources." The program that encourages consumers to choose seafood responsibly was aiming to help develop sustainable consumer behavior gradually.

3.2.3 Housing

The statistics from the Ministry of Construction shows that China completes 1.2 billion square meters of construction every year, 6 times as much as that in Europe, of which, the urban residential construction accounts for 500 million square meters. Large quantities of residential blocks rose up within a relatively short period, and housing has become the largest merchandise for consumption and investment. The consumers are becoming more mature, from simply pursuing bigger floor area to valuing the quality, including location, direction, storeys, greeneries, energy efficiency, water conserving efficiency and property management.

Increasing urban housing space not only leads to consumption of land resources, but increasing consumption of energy, water and construction materials. The remodeling of the old buildings which has gained increasing

---

10 http://www.fao.org/mutrition/education/food-dietary-guidelines/background/sustainability/zh/.

popularity in China, aims to improve the energy conservation. LEED, or Leadership in Energy and Environmental Design, was developed by US Green Building Council (USGBC) at the turn of the new millennium and has become the most widely adopted green building rating system in the world. As a comprehensive system of interrelated standards covering aspects from the design and construction to the maintenance and operation of buildings, LEED has been more and more accepted in China. Between 2008 and 2014, there were 1,410 LEED certified activities related to sustainable constructions, most of which took place in large and medium sized Chinese cities.[11] The purpose of LEED certification is to set comprehensive and precise standards for green buildings, and to avoid the pseudo green concept. Whereas LEED Certification was enforced in some US states and western countries, it was promoted on voluntary basis in China.

In addition, application of new materials and technologies, such as energy conservation, water treatment, household-based heating, solar power and water-efficient toilets, stands for some nice tries in China's urbanization, particularly in the planning and development of eco-cities. In 2014, there were various awareness raising programs promoting sustainable spending on housing, such as sustainable decoration and renovation activities, electrical appliances store's Sustainable Consumption Week, etc.

3.2.4 Transport

Sustainable development in transport, also known as Sustainable Transport or Green Transport, refers to modes of transport with less environmental implications, such as walking, cycling, green vehicles, shared riding and lifestyles that promote and develop urban transport systems. Programs such as high emission old vehicle displacement programs in Beijing, launching of electrical taxis, promotion of P+R car parks and rental bicycles, are examples of low carbon transport means.

Transport system has considerable environmental implications, contributing to 20~25% of the world energy consumption and carbon dioxide emission, including the greenhouse gas emission. Road transport is a key contributor to local air pollution and smog generation. Although sustainable transport is mostly promoted at individual level, it is highly regarded by municipalities, nations, and international communities.

---

11   *Green Building Evaluation Label (China Three Star)*, http://www.gbig.org/buildings/752325.

4   Main Issues, Challenges and Recommendations Regarding China's Sustainable Consumption

The UN Sustainable Consumption China Partnership Summit 2014 was held in a green way. Dr. Nanqing, Jiang, an officer from United Nations Development Programme (UNDP) said, "Since the Partnership came into being in 2012, a series of official and industry promotion activities have been launched, sustainable consumption has gradually driven home to government agencies and the public, more and more Chinese companies came to realize the importance of sustainable production and started to take an active part in the Partnership. Now terms like green consumption, green transport and green buildings, are widely known by the general public, which is remarkable progress and cannot be achieved without the collaborations of our partners."

Nevertheless, there is no denying that China's sustainable consumption is driven by a relatively small number of voluntary consumers, and yet has a long way to go to end the ecological deficit.

The author believes that the main challenges brought by ecological deficit include how to:
(1)   Improve consumption quality;
(2)   Control consumption quantities; and
(3)   Upgrade consumption patterns.

To address these challenges, China aims to reduce the ecological deficit as soon as possible in order to reach the equilibrium between ecological footprint and biocapacity.

The WWF and the CCICED have proposed their respective recommendations. In *China Ecological Footprint Report 2014*, the WWF has provided four major recommendations:[12]
(1)   Improve environment and resource consumption efficiency and promote sustainable consumption;
(2)   Coordinate regional economic development and environmental protection, and guide balanced development of sustainable consumption;
(3)   Control urban and rural ecological footprint, and enable sustainability of urban consumption;
(4)   Encourage public participation in sustainable consumption.

The task force of the CCICED holds that in order to drive successful sustainable consumption, the Chinese government should provide holistic and strategic guidelines. In 2013, the CCICED published *The Strategic Framework and Policy Guideline for Accelerating China's Green Consumption*, which suggests that

---

12   http://www.wwfchina.org/pressdetail.php?id=1539 (Chinese version).

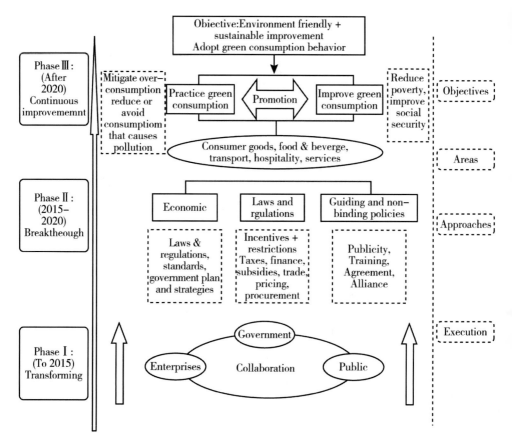

FIGURE 10.1   China's sustainable consumption strategic framework
Note: Http://www.prcee.org/wz/252938.shtml (Chinese version)

China's transforming into green consumption pattern is a state strategy and an integral component of China's overall green development strategic framework, which has specific objectives, key tasks, policies and responsible parties, as shown in the chart above.

Between the recommendations from two organizations, the author prefers the CCICED's, which has taken into full account of the EU's strategic policies on pre-emptive development of sustainable consumption and production, starting certification while the production process is being designed, and complementing sustainable production with sustainable consumption.

The EU has made remarkable progress in promoting sustainable consumption. For instance, organic food market has been growing fast in Europe, accounting for over 50% of the global market in revenue in 2007. Voluntary information tools have been widely adopted, including product ecological labels (ISO1 grade), environmental statements (EPD, ISO111), organic food labels, and consumer awareness raising materials, and so on.

China's urbanization efforts are a huge challenge as well as a great opportunity for sustainable consumption. If we continue to explore sustainable ways for urbanization, shape balanced urban-rural relationships, design sensible urbanization patterns, and promote eco-friendly consumption behaviors, it is likely to bring ecological consumption at a speed lower than urbanization. The examples of some successful ecological cities in the west have demonstrated that even a city is fully developed, the sustainable growth of the economy can further reduce the ecological and carbon footprint.

We can never achieve an ecological and sustainable balance until we fully integrate the sustainable consumption behavior into urban development, and build green production, green lifestyle and green consumption patterns.

CHAPTER 11

The Imminent Threat of Tropical Viruses: Lessons from the 2014 Ebola Outbreak in Africa

*Shen Xiaohui**

Abstract

While tropical viruses can cause lethal diseases, humans are accountable for unleashing them from the tropical forest. It was precisely human activities of killing wildlife and destroying forests that penetrate the ecological barriers of epidemics and thereafter resulted in the spread of diseases across species. Humanity must contain itself before it can truly learn to co-exist with the nature in harmony.

Keywords

Ebola – tropical viruses – ecological barrier – the web of life – biodiversity

In March 2014, new Ebola outbreaks swiped Guinea and quickly spread to other African countries such as Sierra Leone, Liberia, Nigeria, Congo-Kinsasha, Senegal, and Mali. According to the latest statistics released by the World Health Organization, the number of people who died of Ebola worldwide had exceeded 10,000 by the end of March 2015, with a total of 25,000 cases of confirmed infection. Three countries in the West Africa region were stricken the most. Due to the difficulty in collecting data from the vast rural area of Africa, the World Health Organization estimates that the actual number of deaths far exceeds the registered number.

---

* The author Xiaohui Shen is a committee member of the National Committee of the Chinese People's Biosphere and senior engineer of the State Forestry Administration in China. He has long been engaged in the research, writing and environmental protection of protected areas, forests, wetlands, desertification, and wildlife conservation biology. At the time of the Ebola outbreak in 2014, Xiaohui Shen was in West Africa and wrote this article based on his in-depth research of the issue.

This wave of Ebola outbreak was the worst of its kind in history with an accelerating speed of spreading. Many West African countries had to declare a state of emergency. Flights to the epidemic areas were banned and canceled. An ominous air shadowed these tropical African countries as the panic of the virus spread to other countries. Since the epidemic in West Africa was going out of control, the World Health Organization announced that the Ebola outbreak had become a worldwide public emergency and demanded all countries to attach great importance to the development of the pandemic.

1 Three Lethal Viruses from the Tropical Forests in Africa

Viruses are a primitive microscopic organism that is an intermediate form between life and non-life. They are parasites that cannot be "activated" without other cells. A virus is a biological matter yet not completely in life form. It must attack and conquer the host's cells to reproduce itself. This shows that the virus's livelihood is extremely "economical," surviving on resources and energy that entirely come from the host. It is an infective organism that self-replicates with the help of host cells. In the evolutionary processes, viruses generally coexist with the host in peace to obtain a sustainable use of host cell's resources. However, when viruses spread across species, such equilibrium can often be upset. Moreover, hosts from an alien species may also give in completely to the new viruses. Therefore, in such processes, sudden malignant cell lesions can be intensive enough to destroy the host's life.

It is estimated that there are at least 320,000 viruses that can infect mammals worldwide. Historically, humans have been struggling with epidemics caused by viruses. In the era of raging infectious diseases, an outbreak of pandemic can significantly impact historical directions. Some scholars point out that, from an epidemiological perspective, Rome was fallen due to malaria; Egypt was destroyed by schistosomiasis; and the Ming and Qing dynasties in China were weakened by the plague. These pandemics led to social chaos which influenced the fate of a civilization. Now, in order to understand the roots of the three major pathogens that are most harmful to modern society, please allow me to draw your attention to Africa.

1.1 *Malaria*

Malaria is a common human disease that threatens public health in 90 countries around the world. It is a parasitic disease caused by the infection of female Anopheles mosquitoes. More than one-third of the world's population has been infected with malaria. It is estimated that on an annual basis, the number of people suffering from malaria is 300 to 500 million,[1] causing 1.5 to 3.5 million deaths. Most cases of malaria infection and deaths occur in sub-Saharan Africa. During the two years I lived in the West African country Guinea-Bissau which had the highest mortality rate of "brain malaria," I was infected with malaria twice. Fortunately, I was able to restore health after China's medical aid team to Africa provided me with timely and effective treatment.

Scientists analyzed some blood samples from wild chimpanzees in Cameroon and found the true source of malaria parasites. They identified eight species of Plasmodium, and the genes of chimpanzee's Plasmodium is more diverse than that of human Plasmodium, indicating that the malaria is transmitted to humans by the intermediate of the Anopheles mosquito from chimpanzees. Using genetic sequence information, scientists were able to estimate that malaria originated in Africa 65 million years ago, yet it did not spread to areas beyond Africa until 3,300 years ago when it became a global epidemic.

1.2   AIDS

AIDS originates from a virus that attacks human body's immune system. The complex structure of HIV shows that it has undergone an evolutionary process of natural selection. HIV has 9 genes and is so far the most complicated virus known to human knowledge. While the first case of AIDS infection can be traced back to 1959, the first widely-known AIDS case was found in the United States in 1981. In 1985, the disease was officially dubbed the name of "human immunodeficiency virus" (HIV). In Africa, the geographical distribution of the AIDS population closely matches that of a species of mangabeys,[2] yet the infected animals are not killed by the virus. It is said that the hosts of HIV can be traced back to Central Africa in the early 20th century when a hunter entered the tropical forest of Cameroon and killed a chimpanzee. The virus in the chimpanzee's blood entered the human body through the hunter's wounds and changed the fate of human beings. HIV has infected at least 65 million people to date, killing about 2 million people each year, and infecting 6,000 new patients every day. AIDS patients in Africa account for more than 70% of the world.

Scientists have two diverse views regarding the origin of AIDS. One perspective is that HIV was originally hosted by human beings since humans evolved from apes millions of years ago, only was discovered and named in recent decades. However, more scholars believe that HIV is not possessed by humans but is caused by cross-species infection of viruses from chimpanzees. After

---

1 Translator notes: the original article was 30,000 to 500 million which could be a typo.
2 Strictly speaking, mangabeys are monkeys instead of apes claimed by the author.

studying the genes of our ancestors, some American scientists believe that since humans and chimpanzees branched out from the evolutionary tree, the two species had sexual relationships for 4 million years and had common offspring. Many scientists also believe that mankind paid a heavy price for such sexual activities with AIDS being one of the worst consequences. Today, the recurrence of AIDS pandemic is fueled by modernity. It has become a disease of civilization and urban lifestyle, which has created an ideal environment for the activation of an ancient virus to spread globally through airplanes.

1.3 *Ebola*

The Ebola virus is more terrifying than HIV. It is a haunting hemorrhagic plague in Africa. People who have been infected with the Ebola virus rarely survived, with most patients died after one week of initial symptoms. There is still no treatment to cure or prevent Ebola. The deceased victims of the disease have blood coming out of their eyes, ears, and nostrils. Such scenes are particularly chilling. The inner organs of the victims are eventually devoured by the virus, turning into a shapeless slimy paste.

In fact, Ebola is not a new type of infectious disease. It has been popular for centuries in the tropical jungles and grasslands of Central Africa. Ebola had not been well-known because it did not lead to massive deaths previously. The earliest record of the Ebola outbreak was in September 1976, which spread from Sudan to a small town in Zaire called Yambucu (now in Congo-Kinsasha). At the time, this undisclosed yet highly contagious virus ravaged 55 villages and caused 280 deaths. A group of experts from several western countries named the virus after the Ebola River near Yambucu. Since then, there have been more than 30 Ebola outbreaks. Nearly a thousand people died in the 1995 epidemic. Thousands of gorillas also died in Congo and Gabon. After the outbreak of the pandemic, the forest was silent, and the nature was terribly quiet. This time in West Africa, the Ebola eruption was most unusual because it had only happened in the central and eastern parts of Africa. Moreover, none of the incidents had spread as wide as in 2014 with so many cases of infection and death.

The Ebola virus was discovered, but the source of the virus and the route of transmission became a mystery for some time. Ebola is different from diseases like AIDS which causes slow deaths and leaves ample time to reproduce and spread. Ebola uses some animals that carry pathogenies that humans still do not know so well. The reason why the pandemic has not yet caused a pandemic in the world like AIDS and malaria is not that humans have done a perfect job of prevention and control, but because the Ebola virus can kill the victims before it spread. As a result, the virus disappeared with the patient without a trace. As a result, humans found the emergence and disappearance of the disease have been irregular and unpredictable. It is foreseeable that the Ebola epidemic will vanish and recur after hibernation.

Only by locating the origin of the virus and verifying the media and channels of its infection, we could be able to take measures to cut off the origins of the disease and permanently control the epidemic. Scientists believe that the origin of the Ebola virus may be a kind of wild animal that has no direct contact with humans. It is very likely another kind of animal that has direct contact with humans acts as an intermediate host and transmits the virus from the original host to the crowd. This theory is able to explain the sudden appearance and disappearance of Ebola in the past decades. To test such hypothesis, scientists from various countries in Africa have examined mosquitoes, bugs, mice, pigs, cows, monkeys, bats, and deer. In 1996, scientists captured and analyzed 48,000 animals in the Central African Republic alone. However, they didn't find any substantial evidence.

2   A Bane of Disturbing Wildlife

In 2014, after the first outbreak of the Ebola epidemic in four countries in West Africa, 17 European and African experts in tropical diseases, ecology and anthropology formed a research team to investigate and trace the origin of the Ebola virus. They performed animal testing near a small mountain village in remote eastern Guinea, where the first outbreak of Ebola was recorded in December 2013. Scientists discovered that at the time, a two-year-old child was bitten by a fruit bat that had been infected with the Ebola virus. The boy then transmitted the virus to his mother, both of whom died within a week. Later, as the Ebola virus spread with the crowds who came to attend the funeral, it spread farther and farther and expanded.

Previously, although scientists have discovered that bats are the main carriers of the Ebola virus, it was rare to observe cases of infection from bats to humans. The Ebola outbreak was caused, in many cases, by eating infected wild animals. In the Congo rainforest and its surrounding areas, chimpanzee meat is an integral part of the dietary structure of indigenous peoples, and chimpanzees rush to eat young children from time to time. In a sense, it may be said that the food chain relationship in tropical Africa is as old as prehistoric.

In January 1995, a farmer from the outskirts of Kikwit in Zaire died of hemorrhagic fever. The epidemic quickly spread, killing 245 people in 4 months, including 60 medical staff. However, the investigation of the Ebola virus host by scientists discovered nothing. In February 1996, in a small remote village

Mayibout at the junction of Gabon and Zaire, 18 people fell sick at the same time. Before the illness, they were believed to slaughter a chimpanzee who died in the jungle and ate it as a meal. Scientists later noticed that the chimpanzees were dying due to Ebola infection. Subsequently, they also discovered Ebola virus in samples of three large fruit bats. At this point, after more than ten years of field research, scientists finally traced the hosts and intermediate hosts of the Ebola virus.

We have also concerns over the diet of people living in Gabon and Cameroon. Locals eat most kinds of wildlife they hunt. In the bazaar of Congo (Brazzaville), we have seen monkeys being slaughtered and roasted for sale. In Gabon, people chopped off the heads of black colobus and make them into barbecue items; and in Guinea, bat soups and smoked bats were local delicacies; not to mention the common hobby of indigenous people in African jungles: eating chimpanzees and gorillas. Therefore, such trade of "bush meat" has directly led to the spread of waves of various diseases across species to humans. When people are eating these wild animals, the viruses carried in them entered the human body and are in fact engulfing humans too.

Chimpanzees, gorillas, and monkeys are all hosts for the Ebola virus, while bats are intermediate hosts for spreading the virus. In Guinea in March 2014, it was bats that caused a major outbreak of the Ebola virus. Studies have shown that the level of metabolism of bats can increase 15 times when they are flying (compared with birds which only increase 2 times). As their metabolism level soars, the body temperature of bats rises. Increased body temperature can stimulate many immune responses in mammals and produce more antibodies. Bats are able to reduce the toxicity of various viruses in the body through their mechanism of continuous high fever. As a result, bats have become highly-effective repositories of viruses, and their ability to retain mobility while carrying the viruses enables them to become highly-effective spreaders of viruses.

Bats are hosts of many deadly diseases and viruses. It is due to the destruction of their natural habitats, bats have to approach human settlements and live in fruit trees and trees with flourishing branches. Of all the different kinds of bats, the fruit bats are the most dangerous kind. They will leave the virus on the peel and flesh of the fruit through their saliva and urine. People who eat these virus-infected fruits will be infected.

Bats are mammals with an ancient ancestry. Their superior flying capability has enabled them to acquire a vast mobility space and to contact with a large number of plants and animals. This increases the possibilities to spread the pathogens they carry. While bats can carry as many as 61 kinds of viruses that can cause human diseases, they themselves are immune from the symptoms and therefore are particularly scary. Once the viruses are transmitted to humans, they will create a storm of plague that spreads and cause deaths everywhere.

American medical historian Howard Michael, author of *The Story of Pestilences*[3] points out that humans have not yet learned the right way to live in harmony with nature. Some viruses that have been hibernating for thousands of years are constantly being awakened by the enhanced ability of humans to conquer nature. In particular, since the era of agricultural and industrial civilization, the continuous and intense development of the wilderness has exposed humans to more viruses than ever before. Those diseases that were isolated in the depths of the jungles were "discovered" one by one by humans. When humans destroy the ecological barriers that block the disease, the viruses have successfully crossed the boundaries of the species. The war between ethnic groups or countries, as well as population migration, colonization, trade, and tourism, further spread the epidemic diseases to all corners of the human life circle.

It is worth noting that, since the colonial era, the spread of plagues is often across continents as well as across the colonists and indigenous peoples. For instance, smallpox was originally only popular in the Old World, including European, Asian and African continents. However, when European colonists landed in America, they brought many kinds of infectious diseases with them to indigenous people whose immune system was not prepared for those diseases. The most lethal kind was smallpox. The ancient Aztec Empire in the Mexican region suffered from a loss of population from 25 million to 3 million within less than 50 years due to the smallpox plague brought by 300 colonists of Spain. The survivors lost their morale to fight and a powerful empire had fallen. After the plague eradicated the Aztec Empire, it continued to advance southward and also destroyed the Mayan civilization. Due to the smallpox epidemic, the ancient Inca civilization that existed in Peru and neighboring countries today was also easily conquered by 180 Spanish colonists. The European colonists, with or without intention, spread smallpox to the aboriginals in North America. Under the ravages of the smallpox virus, several major aboriginal tribes shrank from millions to only a few thousand people, some even completely died out. Before coming into contact with the colonists, there were 20 million to 30 million inhabitants throughout the Americas. By the 16th century, only 1 million people survived.

---

3 Translator notes: I was not able to find the original English version of the book, neither any historian who has the name Howard Michael. However it was published in Chinese by Shanghai Academy of Social Sciences Press in 2003.

By the mid-17th century, nature took the revenge on the colonists. Slaves from Africa brought malaria and yellow fever to the Americas. Europeans there were apparently more susceptible to these two tropical diseases than local aboriginals.

Today, while AIDS is still rampant and Ebola is surging, the spread of the viruses has been significantly accelerated by modern means of transportation. People who live thousands of miles away across the oceans can be connected by a single flight. In this long-running fight against the viruses, the world will share the same fate. It is hard for any nation or ethnicity to be exempted from the epidemics.

3   We Share the Same Health Prospects across the World

The fight between viruses and humanity can easily make us overlook the other side of the story—microbiological organisms play a vital role in the global ecosystem. It is widely known that the saprophytic bacteria that keep our biosphere clean by decomposing the decaying corpses of plants and animals and releasing essential nitrogen back into the soil for plant growth. Microorganisms transform the energy and material on this planet to keep the vibrancy of their circulation, which sustains the continuity and the health of animals and plants. Without these microorganisms, all types of ecosystems in the world will be suffocated by the junks they produce themselves.

The human body can also be seen as a self-contained "ecosystem." The healthy operation of this ecosystem also depends on the participation of microorganisms. Large volumes of bacteria co-existed in the human body including skin, mouth, and stomach, contributing to daily life activities. The symbiosis of humans and bacteria benefit from the variety and volume of bacteria. Although only a small portion of bacteria is harmful to human health, their destructive power is astonishing. As Nobel Prize laureate Joshua Lederberg[4] said, "Nothing guarantees that we can always win in the natural evolutional competition between viruses and humans." Despite the advancement in science and technology to get out of such race, humans are still far away from becoming a super species that is not constrained by any biological mechanisms. Humans have strived hard to improve their own life quality but nothing fundamental changed. The natural laws are far more powerful compared with human social laws.

---

4   Translator notes: The author misspelled the name of this Nobel laureate and I was not able to find the original version of the quotes.

The existence of the viruses also reveals that the structure of the food chain (or "biological chain" in a more precise term) is not a pyramidal linear one but an interlocking network. The mutual food supply relationships between different species are by no means a simplified unidirectional one as "big fish eats small fish, small fish eats shrimps." In fact, the biological chain consisted of a very complex network which facilitates the circulation of energy, material, and information within this "web of life." This forms an interesting network that features interdependence, mutual constraints, and co-evolution. Thus, it is not accurate to say that humans and predators stay at the top of the pyramid of food chains. Human beings are actually an "unwilling" node in the food chain or ecological chain. They are both "consuming" and "being consumed." Regardless of how greedy mankind seizes natural resources by their hegemony gained over millions of years of evolution, attempting to fill their endless desires, they themselves cannot escape the destiny of being consumed by the decomposers of nature. In fact, they are already slowly "consumed" by bacteria and viruses while they are still alive. This is precisely how mother nature achieves her "fairness" and "justice" through the web of life.

When we learned that the biological chain (including the food chain) is an interlocking network system, and the relationships between organisms are both mutually dependent and restrictive, and all species are equal in a way, we would understand that mankind's attempt to "eliminate" certain species that are "harmful" to the interest of themselves is essentially a short-sighted move to tear the web of life. A temporary solution to the current problem will lead to bigger and more problems. For instance, when people learn that bats are the carriers of Ebola virus and will cause major outbreaks, what measures will be taken? The usual practice is to vigorously eliminate this wild animal that acts as an intermediate host. However, the result of doing so will simply lead to another "ecological disaster"—particularly considering the fact that it is bats that pollinate crops and eat large amounts of mosquitoes. The drastic decrease of the population of bats will inevitably lead to failures of growing crops and controlling mosquitoes, causing famine and the spread of mosquito-mediated tropical diseases such as dengue fever and malaria, etc. Any missing link in the biological chain may cause an unexpected chain reaction. In the face of the epidemic, human beings are more likely to benefit from restraining ourselves instead of attacking others.

In the biosphere, microorganisms including pathogenic viruses are a link between various organisms or different species. Biologists believe that it is the inevitable result of natural selection and evolution. When the ecosystem is relatively stable, the viruses will be lurking in the hosts. Conversely, any disturbance in an ecological perspective, mainly human interference including deforestation and wildlife hunting, can cause the spread of pathogens across species and triggers epidemics.

The tropical rainforest is an unrivaled vast gene bank of diverse species, and possibly also the largest gene bank of viruses since the diversity of the virus is also an integral part of biodiversity. Humans cannot and need not eliminate all viruses. All we need to do is to learn how to co-exist with viruses to maximize their benefits and minimize their harms, so that the virus can, like other living things, benefit all humanity.

CHAPTER 12

Chinese React to Jack Ma's Hunting Trip to UK

*Liu Qin**

Abstract

Do human beings, who are at the top of the food chain, have the right to kill wild animals? After Jack Ma's 2014 hunting trip in Scotland was exposed by the media, his remarks that "hunting promotes wildlife conservation" triggered a heated debate in China. The environmental and ecological consequences, as well as the pusher behind hunting have become the focus of public debate.

Keyword

hunting overseas – celebrity effect – ecological consequences – interest-related pusher

When the Chinese rich entrepreneur Jack Ma's hunting trip in Scotland was exposed by the media in early August 2014, people in China were shocked to find that while some Chinese are still indulging in shopping sprees overseas, the rich have started to yearn for something more exciting. Hunting has become their new hobby.

Hunting is also known as "sport hunting," "trophy hunting" or "game hunting."

The HK media have reported that the Chinese rich pay averagely £100,000 for a hunting trip in the UK. Inspired by the British TV series *Downton Abbey*, these rich people fall in love with hunting in the UK. They would rent castles and hire butlers and staff, and would even wear British aristocratic hunting attire.[1]

---

* Liu Qin, editor of *chinadialogue*, Beijing Office. She has contributed to several articles on environment and ecology.
1 Yang Ningyi, "Hong Kong Media: The Chinese Rich are Splashing Out 1 Million on Hunting Trips to Britain," http://world.cankaoxiaoxi.com/2014/0811/456756.shtml.

However, different from other rich people who "hunt for fun or to show off," Jack Ma believed that he went to Scotland to study hunting in order to know more about the wildlife conservation and the ecosystem.

Feng Yongfeng, founder of the environmental NGO Nature University, criticised Ma in an open letter, saying that it was not worthwhile for him to spend an exorbitant amount of money to learn some so-called "advanced skill" which probably would never be applicable in China.

Jack Ma, when responding to the accusation, said that hunting was a special skill, a way of connecting human beings with nature, and a reflection of the warrior spirit. He said that many countries and regions had raised huge sums of funds for the conservation of nature by organizing hunting and fishing events. However, he also admitted that hunting was extremely cruel.[2]

Jack Ma's response triggered a heated debate between the pro-hunting and against-hunting groups.

The pro side believe that hunting is a scientific and effective way to promote the conservation of wildlife, a fact that the Chinese public choose to ignore. They think Ma's action sets a good example for promoting scientific conservation of wild animals.

The against side, however, claim that Ma has set a bad example for the Chinese rich and will have a negative impact on wildlife conservation. They believe that once the Chinese domestic hunting market starts to open to the rich, it will inevitably undermine wildlife conservation.

1   A Price to Pay for Celebrity Hunting

The public have called Jack Ma a hypocrite, saying that he was talking about environmental protection while at the same time he was hunting.

Celebrities who claim that they are animal lovers have a price to pay when they pick up a rifle and point at an animal. Jack Ma would never have expected that even thousands of miles away in Scotland, he would become the target of public anger the moment he fired his first shot. Looking back, Ma was not the first celebrity who got into trouble with hunting.

In April 2012, King Juan Carlos I from Spain had to make a public apology after his elephant hunting trip in Botswana was exposed by the media. Members of animal conservation organizations expressed their anger and

---

2 Zeng Liang, "Jack Ma Admitted Spending 500,000 yuan for UK Hunting Trip but Claimed No Desire for Aristocratic Lifestyle," http://sports.dzwww.com/china/201408/t20140811_9714029.htm.

organized petitions, and some even called for his abdication. Later the WWF's board in Spain voted unanimously to end the King's job as the institution's honorary president.³

In February 2004, Prince Harry from Britain was involved in a "hunting-gate" scandal after a photo of him with the body of a buffalo he killed on a hunting trip in Argentina surfaced. Just a few days before the incident, he had pledged to save animals threatened with extinction in Africa. His brother Prince William was also exposed by the media for a hunting trip made not long before that. This had aroused public attention and Prince Harry was called a hypocrite whose "acts belie words."⁴

Is it fair to condemn Jack Ma, a self-proclaimed environmental enthusiast, just because of one hunting trip? Xie Yan, a researcher from the Institute of Zoology of Chinese Academy of Sciences, said that as a public figure who is enthusiastic about environmental protection, Ma should have a keener awareness of the current status of environmental protection in China. For a person under the spotlight, his hunting acts would set an example and attract followers, particularly the rich in China.

Ma, nevertheless, is willing to show his passion for environmental protection. He used "Member of Board of Directors of TNC (The Nature Conservancy)" as the only tag for his microblog.⁵

2 The Rich's Hunting Trips are Worsening the Environment in Scotland

Jack Ma didn't expect that his hunting trip not only was reproached by Chinese but poked a sore spot of Scottish environmentalists. The luxurious hunting trips by the Chinese rich did not bring any change to the ecosystem in Scotland, rather they damaged the local environment.

Chinadialogue reported that to attract more clients, the hunting estates in Scotland deliberately allowed the deer population to grow. They killed the natural enemies of the deer to ensure that clients could have enough prey to hunt. The huge number of deer damaged the local environment. Scotland used to have vast expanses of forests, now forest coverage is only 4% with extended

---

3 "The King of Spain was Force to Apologize after an Outcry over His Luxury Hunting Trip and Accidental Injury," http://news.qq.com/a/20120419/000820.htm.
4 Zhao Yanlong, "Prince Harry Caught in the 'Hunting-gate' Scandal: The Photo of Buffalo Hunting Exposed," http://world.huanqiu.com/regions/2014-02/4838311.html.
5 Jack Ma's microblog, http://t.qq.com/tncmayun/.

forest covering only a few kilometres. The hillsides are bare, with no extended natural forests, woodlands and bushes. This is the result of excessive grazing of hundreds and thousands of deer. What's more, the remaining forests are shrinking year by year.

Mike Daniels, Head of Land Management of John Muir Trust told *chinadialogue* that the aristocratic "Downton Abbey" style of hunting was intensifying the environmental problem. Those large hunting estates in Scotland offer luxurious accommodations with private stalkers and butlers to guarantee a lavish lifestyle for their clients. Most of the hunting clubs in Scotland offer this long-practised aristocratic hunting style to attract wealthy clients worldwide. Jack Ma was only one of them.

3   The Invisible Pusher Behind

What the rich have fail to see is that their extravagant consumption has facilitated the increase of deer population in Scotland. Mike Daniels said that it was the Scottish hunting clubs that caused the number of deer remain high in Scotland.

The hunting clubs are inevitably pursuing profits, hence the real pusher behind the hunting is interest not conservation.

A hunting license of a prey often costs tens of thousands of US dollars at international auction. If added the service charge, the cost is even higher, sometimes much higher than the value of the prey itself. Therefore, hunting is deemed as "the game of the rich."

On February 28th, 2012, *Guangzhou Daily* carried a news report entitled "500,000 yuan for a Polar Bear?" The reporter said that from the website of LU Bin, founder of "I Love Hunting Club", the price of hunting trips ranges from 59,800 to 498,800 yuan, covering places in Africa, North America, South America, Oceania and Europe. With the high price come the lavish accommodation, high-end hunting gear, and various tailored services, e.g. private interpreters and stalkers. Different from other luxury tours, a large share of the cost of a hunting trip is on the prey. Take the most expensive 14-day Arctic Polar Bear hunting trip as an example, the exorbitant price of 498,800 yuan includes the price of a boar.[6]

An article entitled "Hunting, a Secretive Game of the Rich" was posted on the "I Love Hunting Club" website. The article cited a report from *The Guardian*, in

---

6   "500,000 yuan for a Polar Bear?" http://ucwap.ifeng.com/tech/discovery/qiquziran/news?aid=32170770&mid=9kgFAI&rt=1&p=2.

which George Goldsmith, head of a firm organizing sports and shooting trips, said that luxury hunting and shooting trips are very expensive and a group can very easily spend £15,000 per day. Nevertheless, the demand has never been so strong. Knight Frank, a UK real estate company also reported that the value of assets related to hunting, shooting and fishing has increased by 32% for the past 10 years. The price of estates with moorfowl has increased by 49%.

China Newsweek reported that to cater to the rich people's interest of collecting ivory, Lu Bin's website would strongly recommend to its members the "680,000 yuan for two elephants" program. "Our elephant hunting trip is legal because we have the Washington Convention import permit. In this case, ivory import into China is legal," Lu Bin said. Zhengan Travel Agency stated on its website, "A 27,655-gram uncarved elephant tusk is priced at 1.68 million yuan, and a 3450-gram tusk at 240,000 yuan in an arts and crafts shop in Beijing. However, with our arranged trip, you can enjoy an unforgettable hunting experience with the luxury of travel and service previously only exclusive to the European colonists in the past centuries. On top of that, you can legally own a pair of 30–80-kilogram elephant tusk and an elephant hide that are highly valued as collectibles.[7]

4 From Hunting in China to Hunting Overseas

According to *China Newsweek*, the person who brought "trophy hunting" to China was a Chinese American named Liu Guolie. He made his fortune in running supermarkets in the US. After he retired in 1974, he started travelling around the world. Since he loved hunting, he made acquaintances with many managers of hunting estates around the world.[8]

Liu came to China in 1984 to promote hunting. Within a year, bharal hunting was started in Dulan County, Qinghai Province. And several other provinces quickly followed suit: argali hunting in Gansu in 1987 and in Buerjin in Xinjiang in 1990; red deer hunting in Sichuan in 1992; and takin hunting in Shanxi in 1993. Liu said that he had helped a number of international hunters hunting in China, "I helped the richest people around the world, among whom are ambassadors, doctors, Arabian princes."

---

7 Xiao Suo, "Chinese Rich Hunting Overseas," http://www.gd.chinanews.com/2012/2012-03-20/2/183120.shtml.
8 Xiao Suo, "Chinese Rich Hunting Overseas," http://www.gd.chinanews.com/2012/2012-03-20/2/183120.shtml.

An insider told *China Newsweek* that the original intention of the forestry department of China to open its wildlife hunting market was to earn foreign currency. No one at that time had ever thought that one day Chinese would hunt overseas.

China's economy has been growing rapidly for the past 20 years, and an increasing number of rich people start to have an interest in hunting. Therefore, Chinese rich people started the hunting trips all around the world. From the African savannah to the freezing North Pole, more and more wild animals have been hunted down by the Chinese rich. Since 2004, an increasing number of Chinese rich have gone outside China for the controversial trophy hunting.

As reported by *China Newsweek*, similar to what happened in the past when foreigners came to hunt in China, now these Chinese rich people have to pay from 50,000 to 680,000 yuan to hunt big games like black bears and elephants and win "trophies" of hides and elephant tusks. Only those who are either rich or in power can afford to the exorbitant price.[9]

In 2014, cases of netizens posting "trophies" online aroused public concern. These people hunted illegally and showed off their experience by uploading pictures to social media. In one case, a man posted a photo of the body of an Equus kiang, a first-class national protected animal in China.[10] In another case, a photo of an ocelot was posted on QQ (a Chinese social media platform) and later reposted by others to Sina Microblog (another Chinese social media platform).[11] Yet in one other case, a person showed off the killing of a white crane by choking the neck of the crane while dragging its wings.[12] The two people in the first two cases finally gave themselves up to the police under the public pressure, while the third case remained unsolved.

Are the people who show off their prey online fearless or ignorant? If fearless, they are in fact taunting the Chinese law enforcement agencies about not having effectively enforced the law. No wonder every year during the migration season, many migratory birds would end up at some Chinese's dinner table. There was even a bizarre case of a few people in Guangxi who liked tiger meat so much that they initiated a group purchase for bargain prices. If we think that those who posted their illegal prey online are ignorant, then much should be

---

9   Xiao Suo, "Chinese Rich Hunting Overseas," http://www.gd.chinanews.com/2012/2012-03-20/2/183120.shtml.
10 "Man Confessed Torturing an Equus Kiang and Dismembered It," http://www.guancha.cn/broken-news/2014_08_14_256666.shtml.
11 Li Xiaojun & Xue Xiaolin, "Guangxi Man in Detention for Killing Ocelot," http://www.thopaper.cn/www/V3/jsp/newsDetail_forward-1283435.
12 Xin Hua, "Netizens Showing Off the Killing of White Crane and Ocelot," http://news.sina.com.cn/c/2014-11-24/090031192467.shtml.

done to popularize the relevant law. In July 2014, a farmer WANG from Henan Province was arrested for killing 87 toads. The court believed that he had violated the hunting law and caused damage to the wildlife resources. According to the report, WANG was probably the first person in China who was arrested and found guilty of catching less than 100 toads.[13]

There is a difference between the Chinese rich hunting legally overseas and people showing off their illegal prey online, yet the nature of these two actions is the same.

A manager of an overseas hunting estate once said that some Chinese rich people liked to hunt as many prey as possible, and would throw away some if they could not take them home. "People are happy to hunt many prey, pose for the photo op and then to show off."[14]

Legal hunting sends a wrong message to the public—money can buy the lives of wild animals. Showing off their prey online reflects some Chinese's distorted value system, i.e., killing a wild animal is not despicable, on the contrary, it is a symbol of wealth and social status. This misconception is so popular that the Chinese government has released several times the regulation on managing business receptions for government agencies, stipulating that business receptions should not offer dishes with protected wild animals such as shark's fins and bird-nests.

5 Can Science Answer the Questions about Hunting?

Is hunting an effective way to wildlife conservation? This is a controversial issue even countries that have quite mature system of hunting cannot easily tackle.

The answer is no according to an article entitled "Will Trophy Hunting Effectively Protect Wildlife?" This article further argues that issues related to hunting may not be easily solved by science. Take the revenue generated by hunting as an example. In most cases, the revenue goes to the organizations that arrange hunting trips, with very little money used for conservation.[15]

---

13 "Farmer in Detention for 3 Months for Killing 87 Toads," http://news.sina.com.cn/c/2014-12-02/005931230425 shtml.
14 Yang Ningyi, "HK Media: Chinese Superrich Paying 1 Million yuan for UK Hunting Trip," http://world.cankaoxiaoxi.com/2014/0811/456756shtml.
15 "Will the Trophy Hunting Effectively Protect Wildlife," http://conservationmagazine.org/2014/01/can-trophy-hunting-reconciled-conservation/.

Theoretically speaking, it is possible to hunt scientifically based on a tight control and close monitoring of the number of animal species. However, this process needs a large sum of research fund which not all countries are able to afford.

The Nature University used the Canadian grizzly bear hunting as a case study. Even those biologists who support bear hunting would agree that it would take 10 years of research and investigation, with about 20 million Canadian dollars to fund the study. Since the Canadian government could not afford the fund, it had to admit that it felt "difficult to defend hunting." If this was the case with Canada, it would be even more unlikely with African countries and China.[16]

What's more, science cannot solve ethical problems. In 2006, when the State Forestry Administration of China announced the rule of "hunting the male not the female (animal); hunting the old not the young (animal)," it quickly became a target of public derision.

Regardless of the question whether human beings have the right to deprive old male animals of their right to live, the horrible experience of the killing of one member of the family would leave a permanent haunting memory to the younger generation. Chris Draper, an expert from Born Free Foundation said that the scene of the prey putting up the last-ditch struggle after being shot would leave other members of the family psychologically traumatized. This was proven true by a case in which an elephant calf killed another calf after the mother elephant had been killed.

Science seems to be even more helpless when it comes to institutions and rules. It cannot offer a solution to the violation of rules and laws. In November 2014, a civil servant in Hunan Province shot and killed a female farmer while hunting. The incident caused a public outcry and the hunting society was criticised for becoming a club for the rich and powerful. A civil servant who should have known better broke the law to indulge a craving for hunting. In such circumstances, a civilian can be easily killed, let alone a wild animal.[17]

---

16 "(An Open Letter) Be Aware of the Chinese Rich's 'Oversea Hunting Syndrome'," http://www.nu.ngo.cn/shsj/1770html.
17 Song Kaixin, "Civil Servant Shot and Killed a Civilian while Hunting," http://news.qq.com/a/20141211/035439htm?tu_biz=1.114.1.0.

CHAPTER 13

Reflections on Outbound Investment by China's Mining Industry

*Bai Yunwen and Bi Lianshan**

Abstract

As a key investment field for China's "going global" strategy, outbound mining investment is facing increasing environmental and social risks. This has led to economic losses and discredit on the part of Chinese enterprises and hardships in the implementation of China's "going global" strategy. Through case studies of China's outbound mining investment projects in countries such as Peru and Laos, this article aims to analyze the problems, identify their causes, and make recommendations on how Chinese enterprises can achieve a win-win situation in their global operations.

Keywords

Chinese outbound investment – mining investment – green credit – NGO participation

With a fast-growing economy and accelerating "going global" strategy, China has exerted remarkable influence with its outbound investment. According the Ministry of Commerce's 2014 statistics, China's foreign direct investment (FDI) reached USD107.84 billion in 2013, exceeding the 100 billion benchmark for the first time.[1] China has been Africa's biggest trade partner since 2009 and became the third largest investor in Latin America in 2010. Outbound investment has become an important source for China's economic growth. China's

---

* Bai Yunwen is Director and Researcher at the Greenovation Hub. She has years of research experience in sustainable finance, outbound investment and climate and energy policies. Bi Lianshan is a research specialist at the Greenovation Hub. She is specialized in outbound investment and has participated in an onsite survey in the Great Britain and East Africa on the impact of China's outbound investment.

1 Ministry of Commerce of the People's Republic of China, National Bureau of Statistics of China, and State Administration of Foreign Exchange, *2013 Statistical Bulletin of China's Outbound Direct Foreign Investment* (2014), 3.

overseas engineering projects center on electricity, mining, road and rail construction, and infrastructure related to people's livelihood. These projects boost local economic development.

Yet due to inadequate understanding of the host country's cultural, political and legal environment, some investment projects have experienced resistance from the host country and local communities. This is a challenge that many Chinese enterprises face in performing corporate social and environmental responsibility. Examples of local resistance include the Myitsone Dam, which was discontinued by Myanmar government in 2011, and a road construction project in Sri Lanka, which was suspended in 2015 due to the absence of environmental impact assessment.[2] Environmental issues often give rise to serious social risks in China's outbound investment, leading to economic losses and discredit on the part of Chinese enterprises and controversy over China's "going global" strategy.

The mining industry highlights the opportunities and potential risks brought by an outward-looking economy. With the introduction of China's "going global" strategy in 2000, more and more mining companies started to look for business opportunities overseas. At first, enterprises were searching for mineral resources to satisfy increasing domestic demand. Nowadays, outbound investment is an integral part to the enterprises' globalization and diversification schemes as well as a channel to ease employment pressure in China. Outbound mining investment accounts for a large percentage of China's outbound FDI in terms of both investment flow and accumulated stock. In 2013, outbound FDI in mining reached USD 24.81 billion, accounting for 23% of China's total FDI. Mining is one of the top industries of China's FDI, second only to holding investment in lease and commercial services. Chinese mining enterprises are investing in many parts of the world. They take advantage of diversified investment methods, including setting up wholly-owned subsidiaries or joint ventures with local companies, green field investment, equity merger and acquisition, and providing equipment and services for projects by contract. China's outbound mining investment benefits host countries economically, creates job opportunities, and introduces new technology. In addition, it increases people's general income in mining zones and improves local public services such as medicine and education. In 2013, China's outbound mining investment paid USD 37 billion taxes to the host countries and

---

2 Ranga Sirilal and Shihar Aneez, "Sri Lanka threatens Chinese firm with legal action to stop project," *Reuters*, March 5, 2015, http://in.reuters.com/article/2015/03/05sri-lanka-china-portcity-idINKBN0M01Y720150305.

created 967,000 jobs.[3] In 2011, China's investment in Zambia created 50,000 jobs, most of which were in the mining industry.[4]

However, mining has always been a high-risk industry for investment. Environmental and social risks abound in implementing mining projects, including water and soil pollution, production safety, labor, land expropriation and resettlement of affected residents. If these issues are not properly handled, conflicts would arise. For example, when Zijin Mining was acquiring the Rio Blanco copper mine and later when the project was about to start, multiple violent incidents broke out due to disputes over land rights and inadequate communication with the local communities. As a result, the project had to be shelved.[5] Wanbao Mining found itself in a similar situation. Community relations were precarious before it took over Mymar's Letpadaung copper mine. When Wanbao Mining announced expansion plans of the copper mine in 2012, issues of land compensation and waste water from mining led to protests and injuries.[6]

The international community holds different perspectives on China's outbound investment. Investment in mining, due to its high risk, produces more questioning and critical news coverage. Politicians sometimes join the debate and express concerns over China's acquisition of resources. President of Botswana once commented openly, "We have had some unpleasant experience with some Chinese enterprises.... We will closely review every Chinese company, no matter what it engages in."[7] In some cases, criticism harbors political goals. Chinese enterprises that invest overseas have different background and diverse investment methods. Political and social situations in the host country and region are complex. Companies have different track records and corporate governance systems. Because of these complexities, the following case studies may hardly paint a complete picture of China's outbound FDI, but they do highlight, to various degrees, the issues that China's mining

---

3 Ministry of Commerce of the People's Republic of China, National Bureau of Statistics of China, and State Administration of Foreign Exchange, *2013 Statistical Bulletin of China's Outbound Direct Foreign Investment* (2014), 6.
4 "Zambia Seizes Control of Chinese-Owned Mine Amid Safety Fears," BBC News, February 20, 2013, http://www.bbc.co.uk/news/business-21520478.
5 Greenovation Hub, *China's Mining Investment: Development, Impact and Monitoring at Home and Abroad*, http://www.ghub.org/wp-content/uploads/2014/11/PDF2-Mining_ZH_CASE.pdf.
6 Lucy Ash, "Burma Learns How to Protest—Against Chinese Investors," BBC News, January 24, 2013, http://www.bbc.co.uk/news/magazine-21028931.
7 Nicholas Kotch, "Khama Wants Fewer Chinese Firms to Receive State Contracts," *Business Day*, February 20, 2013, http://www.bdlive.co.za/world/africa/2013/02/20/news-analysis-khamawants-fewer-chinese-firms-to-receive-state-contracts.

enterprises face in their global operations and that impact the overall image of China's "going global" strategy. In fact, these issues are not unique to mining. Wherever the investment is, setting a high standard for itself and performing its environmental and social responsibility are the best strategies to cope with high risks. As a researcher at the Ministry of Commerce explained, "Chinese enterprises have resources and technology, but they lack respect for the local culture and underperform their corporate social responsibility. If these issues are not addressed, China's FDI cannot sustain its growth."[8]

To further understand China's outbound investment in mining, a research team from Greenovation Hub, a Chinese indigenous environmental NGO, conducted a case analysis of Chinese enterprises in countries such as Laos, Cambodia and Peru. Known for its local research with a global perspective, Greenovation Hub supports the effective formulation and implementation of environmental and climate policies, facilitates multi-stakeholder dialogue and promotes changes. Greenovation Hub's substantial onsite survey paints a vivid picture of the performance of Chinese enterprises in foreign countries.

1   High Standards Lead to a Win-Win Situation

Chinese mining enterprises have been penalized a number of times for violating environmental regulations. For instance, in March 2014, due to water pollution caused by inadequate waste water handling system, the operation of Aluminum Corp of China (Chinalco)'s Toromocho copper mine was halted by Peru's Environmental Evaluation and Regulation Bureau under the Ministry of Environment. Chinalco was ordered to rectify its practice. During the two weeks of suspension, Toromocho copper mine's production suffered great losses and the entire project was delayed. Ironically, Chinalco had spent a sizeable amount on environmental and social compensation. The acquisition of Toromocho copper mine was completed in May 2008, but operation did not restart until the end of 2013, following five years of pre-operation construction and community compensation. The total costs of improving community relations and environmental protection amounted to USD1.5 billion, whereas mining expenses totaled USD800 million.[9] Chinalco's measures, such as building a new town specifically to resettle the original residents of the mining

---

8   Ding Qingfen, "Chinese Firms' Growing ODI Offers World Opportunities," *China Daily*, July 10, 2012, http://www.chinadaily.com.cn/cndy/2012-07/10/content-15563185.htm.
9   "Why Chinese Enterprises Refrain from any Attempt of Pollution in Peru," http://view.news.qq.com/original/intouchtoday/n2750.html.

zones and compensating these communities, were applauded. Yet, the suspension of operation, caused by water pollution, came only four months after the mine was restarted.[10] Noncompliance with environmental regulations proved to be very costly. Zijin Mining-invested Rio Blanco was fined in 2008 for its failure to follow the restrictions outlined in the Environmental Assessment Report, which its previous mother company had submitted to the Peruvian government and was subsequently approved. In addition, pollution hazards and land ownership issues which existed earlier were not properly settled. All of the factors led to resistance from NGOs and local residents and progress was hard and slow.[11]

Enterprises can cope with environmental and social risks in overseas projects if they follow high industry standards. For example, China Minmetals Corporation purchased the majority of assets of an Australian company, OZ Minerals, in 2009 and established a subsidiary, Minerals and Metals Group (MMG). MMG's assets included the Sepon copper-gold mine in Laos. MMG not only followed the optimized management system of its predecessor but also voluntarily joined the International Council on Mining & Minerals (ICMM) and the Extractive Industries Transparency Initiative (EITI). By adopting these international guidelines, MMG formulated its production management requirements and improved its relationship with the local mining community.[12] As a result, MMG never had any pollution accidents and established trust with the local community. The success of MMG shows that Chinese enterprises adopting high standards in outbound investment can not only reduce or avoid potential investment risks but also promote friendly relationship with the local communities and boost their corporate image.

In order to avoid environmental and social risks, the Chinese government has provided policy guidelines on environmental and social standards for outbound mining companies. For example, the Ministry of Commerce and the Ministry of Environmental Protection issued the "Guidelines for Environmental Protection in Foreign Investment and Cooperation" in 2013, urging overseas

---

10   Zhou Zhou, "China Aluminum Failed to Fully Consider the Impact of the Rainy Season in South America before Restarting Copper Mine in Peru," http://www.nbd.com.cn/articles/2014-04-15/826046.html.
11   Greenovation Hub, *China's Mining Investment: Development, Impact and Monitoring at Home and Abroad*, http://www.ghub.org/wp-content/uploads/2014/11/PDF2-Mining_ZH_CASE.pdf.
12   Greenovation Hub, *China's Mining Investment: Development, Impact and Monitoring at Home and Abroad*, http://www.ghub.org/wp-content/uploads/2014/11/PDF2-Mining_ZH_CASE.pdf.

companies to respect the customs and religious beliefs of host countries, comply with local laws and regulations, and promote the concerted development of the local economy, environment and communities. The Guidelines also put forward requirements on environmental assessment, pollutant control and emergency response plan.[13] China Chamber of Commerce of Metals, Minerals and Chemicals Importers and Exporters issued the "Guidelines for Social Responsibility in Outbound Mining Investments" at the end of 2014. Incorporating international and China's domestic policies, the Guidelines put forward seven guiding principles, including respecting natural resources and various stakeholders.[14] Although these guidelines are not mandatory, the adoption of such standards will help companies and relevant financial institutions improve their ability to prevent and control risks in their investment. In countries and regions where laws and regulations are lagging but people have a strong sense of rights, complying with only local laws and regulations as the minimum standard does not guarantee a project's smooth execution. Since these guidelines have limited binding power on enterprises, further standards are needed for implementation. For example, China Chamber of Commerce of Metals, Minerals and Chemicals Importers and Exporters is conducting training programs on the implementation of the "Guidelines for Social Responsibility in Outbound Mining Investments."

To develop and improve Chinese policies and guidelines, policymakers and enterprises can reference established international standards. The United Nations Global Compact and the Global Reporting Initiative are two examples of the standards that enterprises and banks can voluntarily implement. Some Chinese companies have already participated in these global initiatives. For example, China Development Bank has joined the UN Global Compact, and the subsidiaries of China Minmetals Group have joined the Global Reporting Initiative. This will not only help outbound mining enterprises mitigate risks in their projects and achieve win-win development with local communities, but also support Chinese enterprises in their efforts to adopt a green and sustainable development model.

---

13   Ministry of Environmental Protection and Ministry of Commerce, "Guidelines for Environmental Protection in Foreign Investment and Cooperation" (2013).
14   China Chamber of Commerce of Metals, Minerals and Chemicals Importers and Exporters, "Guidelines for Social Responsibility in Outbound Mining Investments," http://www.cccmc.org.cn/docs/2014-10/20141029161135692190.pdf.

2 Information Disclosure is Critical to Enterprises and Local Communities

Adequate information disclosure is not only stipulated in China's Environmental Protection Law, but also a key concern of local residents and stakeholders in overseas mining investment. Particularly, in large-scale projects such as mining and infrastructure construction, issues such as resource taxes, land rights and resettlement compensation may arise in addition to the already complicated process of project construction. Probabilities of corruption and misuse of funding are high. If perceived as withholding information, companies are likely to be mistrusted by the public and NGOs, thus impacting companies and their projects negatively.

Zijin Mining has a poor record in information disclosure regarding its domestic projects. Zijin had two serious pollution accidents in 2010. Due to its withholding of information regarding the water pollution accident, the domestic water supply in the downstream was affected and fishery industry suffered loss of income. As a listed company, it also hurt the interest of its shareholders. Zijin was investigated and penalized three times by the Securities Regulatory Commission.[15] With its overseas projects, Zijin does not perform well in information disclosure either. Peru's Rio Blanco copper mine is a case in point. On the one hand, the project's pre-acquisition due diligence neglected existing tensions with the local community. After Zijin's takeover, these tensions escalated into conflicts, threatening the project and employee safety. These operational and financial risks were not disclosed to the shareholders and investors in a timely manner. On the other hand, Zijin Mining did not have transparent communication when dealing with local communities on issues such as resettlement and compensation. The lack of trust escalated into conflicts and made it difficult to carry out the project.[16]

Domestic regulations on the extractive industry's information disclosure are less than rigorous. At present, only companies listed on the Shanghai Stock Exchange and the Hong Kong Stock Exchange must disclose information on resource transaction prices. Lack of information disclosure is not unusual. In 2013, Transparency International, an international NGO that

---

15 Greenovation Hub, *China's Mining Investment: Development, Impact and Monitoring at Home and Abroad*, http://www.ghub.org/wp-content/uploads/2014/11/PDF2-Mining_ZH_CASE.pdf.

16 Greenovation Hub, *China's Mining Investment: Development, Impact and Monitoring at Home and Abroad*, http://www.ghub.org/wp-content/uploads/2014/11/PDF2-Mining_ZH_CASE.pdf.

monitors embezzlement and corruption, published a ranking of countries and enterprises in emerging economies based on their anti-corruption and transparency practice. Among them, Chinese companies performed poorly, with an average score of 2 out of 10, the lowest among the five BRICS countries. In the assessment for companies, 5 of the 33 Chinese companies selected were in the extractive industry and all of them were underperforming, among which Yanzhou Coal Co., Ltd. scored the highest with 2.8 points and China Minmetals scored the lowest with 0.8 points.[17] The assessment evaluated general corporate activities rather than overseas projects only. However, the low scores of Chinese companies indicated the need to establish a more transparent corporate culture. Lack of transparency will affect both domestic and overseas projects and will harm the overall image of Chinese companies. Chinese companies still have much room for improvement in this respect.

The Extractive Industries Transparency Initiative (EITI) sets a good example for improving transparency and governance. EITI is a global standard to promote the open and accountable management of oil, gas and mineral resources. In each implementing country, a multi-stakeholder group comprised of representatives from government, companies and civil society is established to oversee EITI implementation. So far 31 countries are EITI compliant. EITI encourages implementing countries and companies to disclose their revenue and taxes related to natural resources. Peru, Mongolia, Nigeria, where Chinese mining companies have overseas projects, are EITI implementing countries.[18] At present, China is not yet an implementing country, but Chinese companies can become its members. For example, some of China Minmetals' subsidiaries have joined the initiative. Chinese enterprises adopting international systems in overseas projects will in turn improve the level of domestic information disclosure.

No matter which country Chinese mining enterprises wish to invest in, it is critical to ensure transparency in the entire process, from the acquisition of mining lease to mining operations. Particularly, in countries and regions with underdeveloped regulatory bodies, weak rule of law, and high corruption risks, adequate information disclosure mechanism and commitments can avoid suspicions from related parties. The relationship with local communities is significant for the smooth operation of the company. We recommend that companies and financial institutions conduct due diligence and communicate

---

17   Greenovation Hub, *China's Mining Investment: Development, Impact and Monitoring at Home and Abroad*, http://www.ghub.org/wp-content/uploads/2014/11/PDF2-Mining_ZH_CASE.pdf.
18   "EITI Country Implementation Status," http://eiti.org/countries.

with local stakeholders, particularly the public, before investing in a project and especially when unfamiliar with the conditions of the host country. Due diligence and adequate communication would help companies make more informed assessments and decisions.

3   Financial Institutions Act as Gatekeepers for Mining Investment

In sectors with significant environmental and social impacts such as mining and infrastructure, projects usually have a large investment scale and a long life-cycle and require financing from commercial banks or development financial institutions. Globally, development finance not only provides important financial support in investment and construction but also establishes governance mechanisms and safety and security policies, which play a key role in ensuring positive environmental and social impact. The financial approach to controlling a project's environmental and social impact will cut off loans to law-breaching companies and therefore raise the cost of environmental violations. More importantly, when effectively including environmental and social security standards in their credit terms, financial institutions encourage loan-seeking companies to proactively prevent financial risks caused by environmental and social issues.

In recent years, Chinese banks have often demonstrated their ability in monitoring corporate environmental behavior through loan management. In 2014, Xuzhou Environmental Protection Bureau shared with the People's Bank of China the annual environmental behavior rating of the enterprises in the jurisdiction, which was incorporated into the bank's corporate credit information system.[19] In 2010, loans to 46 polluting enterprises in Liaoning Province were withdrawn or discontinued.[20] In 2007, upon the State Environmental Protection Administration's report to the People's Bank of China and the China Banking Regulatory Commission, loans to 12 polluting enterprises were withdrawn or discontinued.[21]

At present, China's financial industry is still optimizing its environmental and social security policies and has much to learn from international financial institutions. In order to help financial corporations improve their

---

19 "Xuzhou Polluting Enterprises Saw Existing Loans Restricted and New Loans Prohibited," http://www.025ct.com/xuzhou/xzqy/2014/0610/325609.html.
20 "Loans to 46 Heavily Polluting Enterprises in Liaoning Province Discontinued: Finance to Curb Environmental Violation," http://finance.qq.com/a/20100816/001172.htm.
21 Liu Shixin, "12 Heavily Polluting Enterprises Experienced Loan Rejection by Banks," http://business.sohu.com/20071116/n253288139.shtml.

environmental and social responsibility, the China Banking Regulatory Commission issued the "Green Credit Guidelines" in 2012, which provided guidance to the banking industry on the compliance of loan projects.[22] The Guidelines clarified the risk management of overseas investment projects in Article 21 and provided support for financial institutions to play a regulatory role in a project's environmental and social impact. Article 21 stipulates that China's banking and financial institutions should strengthen the management of environmental and social risks of overseas projects to keep up with good international practices. China's development financial institutions, such as China Development Bank, have taken active steps to improve the environmental and social standards of their investment portfolios and have developed related environmental policies and internal performance indicators based on the ten principles of the UN Global Compact, such as the principles on human rights, environment, labor and corruption. The Export-Import Bank of China has similar policies. Although China Development Bank released an abstract and touched upon these policies in its corporate social responsibility report, it did not disclose the full content of the policy. In 2008, the Export-Import Bank of China issued environmental and social impact assessment policies for the World Bank energy efficiency projects. It put forward requirements for the borrowers to communicate with the community prior to the start of their projects and to minimize environmental and social impacts during the operations. However, the policy was accessible to the public only for a short period of time. Moreover, the bank's assessment of the proposed credit project was conducted behind closed doors, and it did not disclose the specific assessment reports for the approved projects. China Development Bank and the Export-Import Bank of China's funding projects currently do not have a clear channel for public participation or access to information, or a mechanism for the public to file complaints against a project. These practices will compromise the actual effect of green credit policies.

There are many international experiences to draw from to formulate a better green credit policy. For example, the World Bank and the Asian Development Bank already have relatively established environmental and social security policies and have a complaint mechanism if the loan receivers breach any regulations. Thus, development finance can play a regulatory role prior to and after the grant of loan. Chinese banks have joined a number of voluntary international frameworks such as the Equator Principles and the United Nations Environment Program—Finance Initiative (UNEP FI). Seventy-eight financial

---

22   China Banking Regulatory Commission, "Green Credit Guidelines," http://www.cbrc.gov.cn/chinese/home/docView/127DE230BC31468B9329EFB01AF78BD4.html.

institutions in 35 countries have adopted the Equator Principles by 2013. UNEP FI is a voluntary framework with a mission to promote sustainable development. These non-binding, voluntary frameworks can help banks develop internal policies on environmental and social impact for project financing. The Equator Principles provide a minimum standard for due diligence to support responsible risk decision-making. Financial institutions that adopt the Equator Principles commit to implementing the principles in their internal environmental and social policies, procedures and standards for financing projects and will not provide project finance or project-related corporate loans to projects where the client will not or is unable to comply with the Equator Principles. Although the Industrial Bank Co. Ltd. is the only Chinese commercial bank that has adopted the Equator Principles, some Chinese banks have organized working groups to analyze the Equator Principles and to incorporate these principles into their practice. For example, in 2008, the China Development Bank established an Equator Principles working group and stated in the 2012 annual report that these principles would gradually be adopted in the development of the banking industry. Chinese financial institutions such as the China Development Bank, China Merchants Bank, Industrial Bank, and Shenzhen Development Bank have signed up to join the UNEP FI and have signed declarations covering areas such as sustainable development, sustainable development management, and public awareness and exchanges.

4 NGOs Play a Critical Role in Reducing Ecological Footprints in Outbound Investment

China's key role in international affairs places higher demands on the participation of Chinese NGOs. In outbound investment, on the one hand, Chinese NGOs need to have the capacity to participate in international issues, expand their horizon, take into consideration wider concerns than just China's own economic, social, and environmental issues, and serve as a bridge between China and the project's host country. On the other hand, NGOs need to have the ability to promote communication with companies, eliminate misunderstandings, and enable companies to trust their advice on local conditions and feedback on relevant environmental and social impacts. In this process, NGOs can facilitate enterprises and the government's communication with local communities and NGOs, help Chinese enterprises adopt the best international practice when investing overseas and integrate with the world's advanced models.

After analyzing cases of China's overseas investment, the Report on Chinese Enterprises Globalization (2014) pointed out that Chinese companies needed to improve their ability to handle complex political and social relations in the host country, including learning to listen to "the voices of the opposition, NGOs and the media." Chinese companies often take lightly the role of NGOs in FDI. They also place more emphasis on relationships with high-level government officials than with local stakeholders. This intensifies environmental and social conflicts that could otherwise have better solutions.[23] Enterprises should not regard NGOs as an enemy or a scourge but should regard it as a bridge to help them communicate with local stakeholders, source local information, understand local laws and regulations, and listen to the voice of the local community.

Development finance is one area where NGOs play a key role. The development of security policies for multilateral development banks, such as the World Bank and the Asian Development Bank, is inseparable from the participation of and supervision by international civil organizations for more than a decade. International Rivers, an international NGO striving to protect rivers and communities in river basins, has been paying close attention to the financing of hydropower. For the past ten years, NGO Forum on ADB, an NGO alliance with a focus on the social impact of the ADB credit projects, has been monitoring the development of the ADB's environmental and social standards. Through the participation and suggestions of international and local NGOs, safety and security policies can more effectively consider the impact of the project on the local environment and community. Financial institutions have recognized and adopted more recommendations from NGOs. The environmental and social impact policies of the leading Chinese development financial institutions need to be formulated and supervised with the assistance of civil organizations that understand China's national conditions. Although monitoring measures on the environmental and social impact are being worked on, China's development finance has a lot to cover before it reaches international standards. In the past two years, there has been much discussion about China's outbound investment within the circle of Chinese NGOs, but the supervision of development finance has been limited due to the NGOs' limited knowledge and capability on this issue.

Chinese NGOs can play a role in monitoring the operation of Chinese enterprises. Supervision is not intended to limit the investment and development of Chinese enterprises. Rather, it seeks to minimize business risks caused by

---

23   Wang Huiyao, et al., *Report on Chinese Enterprises Globalization 2014* (Social Science Academic Press, 2014), 26.

the company's environmental and social impact while protecting the interests of local communities. More importantly, Chinese NGOs' "going global" is coupled with "introducing" new ideas. NGOs can learn how other countries have incorporated public participation into decision-making, and how government regulations can be promoted to benefit the sustainable development of communities and the environment. Thus, Chinese NGOs can accumulate experience to promote China's economic transformation.

5   Conclusion and Outlook

Through finance, trade and outbound investment, China is rapidly becoming a significant player in the global economy. While adapting to global rules, China also shoulders the responsibility to formulate new ones. Enhancing the image of Chinese enterprises abroad not only benefits the enterprise itself but also supports China's "going global" strategy. Implementing international standards in overseas projects also improves domestic laws and regulations and industry standards, particularly in terms of social welfare and environmental protection. In recent years, China's ministries and commissions are striving to improve policy guidelines, regulate the market behavior of companies and put forward voluntary frameworks. However, in addition to complying with the guidelines and frameworks, enterprises need to adopt new lines of thinking. They need to be more inclusive and open to the concerns of the people in different economic and political systems. Improving infrastructure and creating jobs are important, but more needs to be done to satisfy the host country's appeal for "inclusive development." Returning to China from abroad, these enterprises are in a unique position to apply their rich overseas experience to a domestic setting.

CHAPTER 14

A Case for Banning Illegal Timber Imports in China

Yi Yimin*

Abstract

Illegal logging and related trade have posed serious challenges to regional development, national stability, climate change and biodiversity. In recent years, the issue of illegal logging and illegal timber trade has drawn more and more attention from all parties. In addition to the continuous efforts of timber producing countries to improve their forest management and governance capacity, consumer countries are also taking active actions to combat illegal logging and trade. As the world's largest importer of forest products, China has issued voluntary guidelines to regulate enterprises operating abroad, in order to reduce the negative environmental and social impact of its huge demand for forest products on timber producing countries. On the other hand, China could do more to deal with illegal logging and related trade, including banning illegal timber from entering China with strict legislation.

Keywords

illegal logging – timber trade – tropical forests – environmental crimes – anti-corruption

1   Introduction

In recent years, illegal logging and related trade have hindered the development of timber producing regions and countries, damaged the environment, neutralized global efforts to combat climate change, spawned a great deal of corruption and crimes, and in some cases even contributed to regional conflicts. Therefore, illegal logging and illegal timber trade have drawn attention from all parties. In addition to the continuous efforts of timber producing countries to

* Yi Yimin is the China project consultant of the nonprofit organization Global Witness and life member of Friends of Nature, who focuses on the social and environmental impacts of economic growth on developing countries and the governance model for natural resources.

improve their forest management and governance capacity, consumer countries are also taking active actions to combat illegal logging and trade. The United States, the European Union, Australia and other major consumer countries and regions of forest products have issued related laws prohibiting the import of illegal timber. Among them, the EU Timber Regulation, passed in 2010, was implemented in 2013, which has imposed more strict requirements on the legitimacy of some of China's timber trading and processing enterprises exporting to the EU.

As the world's largest importer of forest products, China is also exploring how to take more responsibility in this regard. A Guide on Sustainable Overseas Trade and Investment of Forest Products by Chinese Enterprises, being made by the State Forestry Administration of China, is one of such attempts. However, the guide is only applied to Chinese enterprises operating abroad. Although it can help Chinese enterprises to improve their overseas operations, it does not really deal with the problem of large quantities of illegal timber entering China through trade. China could do more to deal with illegal logging and related trade, including banning illegal timber from entering China with strict legislation.

2   The Importance of Addressing Illegal Logging and Timber Trade

Forest resources concern the sustainable development of the world, especially for the countries where the resources are located. A large number of poor people depend on forest resources for a living; the soundness of forest governance and timber trading system has a deep impact on the mode of national and global governance and supply chain; forest resources protect biodiversity and effectively address the challenge posed by climate change to the world. However, illegal logging and trade have posed unprecedented challenges to the sustainable management of forest resources. Some of the world's most valuable forests have been devastated, with serious environmental, economic and social consequences. According to the estimates of the Australian government, illegal logging and illegal timber trade bring US$46 billion of economic losses globally per year, while social and environmental losses amount to US$605 billion per year.[1] The major impact of illegal logging and trade at the regional, national and international levels includes the following aspects.

---

1 See Illegal Logging Prohibition Bill 2012, "Revised Explanatory Memorandum issued by the Australian Parliament: 2.2."

1.1 Affecting Communities that Live on Forest Resources

According to the World Bank, 1.6 billion people depend to various degrees on forests and their environment.[2] Of them, about 60 million indigenous people, living in or near dense forests, depend almost entirely on forests for their livelihood. The number of people who are highly dependent on forests for their livelihood and income is about 352 million.[3] In some cases, while illegal logging will bring some financial return to a small number of people in the community in the short term, once the forest is destroyed by illegal logging, the long-term development of the forest industry will be restricted, and the original livelihood of the majority of people in the community will be threatened, which will lead them into food insecurity and extreme poverty, and the culture of the original community will also be destroyed due to the changes in the environment.

1.2 Threatening National Security and Regional Stability

Illegal logging and serious corruption often go hand in hand, most of which are actually "white-collar crimes," where the businessmen obtain documents signed by senior government officials for illegal logging.[4] It can be highly detrimental to the ability of governance and law enforcement of timber producing countries.

When a country's main economic industries are controlled by criminal gangs and flooded with corruption, its economic stability and long-term development will be greatly restricted. Illegal timber logging and trade is a typical example. It is a multibillion-dollar business[5] that breeds global corruption and crime, especially in the Asia-Pacific region. According to the United Nations,

---

2 World Bank, "Sustaining Forest-A Development Strategy," World Bank, 2004, p.16, http://www.WDS.Worldbank.org/external/default/wdscontentserver/wdsp/Ib/2004/07/28/000009486_20040728090355/rendered/PDF/297040v.1.PDF.

3 World Bank, "Sustaining Forest-A Development Strategy," World Bank, 2004, p.16, http://www.WDS.Worldbank.org/external/default/wdscontentserver/wdsp/Ib/2004/07/28/000009486_20040728090355/rendered/PDF/297040v.1.PDF.

4 Illegal logging with government approval exists in a number of timber producing countries, including: Liberia, see Global Witness report "Logging in the Shadows: How Vested Interests Abuse Shadow Permits to Evade Forest Sector Reforms," April, 2013; Malaysia, see Global Witness report "An Industry Unchecked: Japan's Extensive Business with Companies Involved in Illegal and Destructive Logging in the Last Rainforests of Malaysia,, September, 2013; Cambodia, see Global Witness report "Cambodia's Family Trees," June 2007.

5 Nellemann, C, INTERPOL Environmental Crime Programme (eds.), "Green Carbon, Black Trade: Illegal Logging, Tax Fraud and Laundering in the World's Tropical Forests: A Rapid Response Assessment," INTERPOL and UN Environment Programme, 2012, p.6. INTERPOL estimates the economic value of illegal logging globally as being between US$30–100 billion.

about 70% of international trade in illegal timber comes from the region,[6] and illegal timber trade is the second largest source of finance for criminal groups in the region.[7] As organized crimes have increased, Interpol has also taken note of related crimes, such as murder, violence and crimes against native forest dwellers.[8] The Global Witness report in November 2014 also noted that in Peru, the number of murdered environmentalists is rising, including leaders of indigenous communities who opposed illegal logging.[9]

Illegally logged timber has also provided support for many militias and other non-governmental armed groups. Similar to conflicts related to mineral deposits, these groups raise funds by exploiting natural resources, such as minerals and timber, so as to escalate the conflict. Such a link can be found in many unstable regions. In the Democratic Republic of Congo, revenues from logging, charcoal production, charcoal taxes and other sources are used to finance the conflict with its neighboring country; in Liberia, former President Charles Taylor took timber as the key resource for funding at all stages of the civil war; in Asia, timber revenues financed the Khmer Rouge in Cambodia and played a role in the conflict in Myanmar.[10] In this sense, only by banning illegal timber can we contain the growth of related conflicts once and for all.

---

6   UN Office on Drugs and Crime (UNODC), "Transnational Organized Crime in East Asia and the Pacific: A Threat Assessment," UN Office on Drugs and Crime, April 2013, p.96, http://www.unodc.org/documents/data-and-analysis/studies/TOCTA_EAP_web.pdf.
7   UN Office on Drugs and Crime (UNODC), "East Asia Pacific Transnational Organized Crime Flows Generate USD 90 Billion Annually," UN Office on Drugs and Crime, April 16, 2013, https://www.unodc.org/documents/southeastasiaandpacific//2013/04/tocta/Press_release_UNODC_Transnational_Organized_Crime_EAP_16_April_2013.pdf; UN Office on Drugs and Crime (UNODC), "Transnational Organized Crime in East Asia and the Pacific: A Threat Assessment," UN Office on Drugs and Crime, April 2013, p.96, http://www.unodc.org/documents/data-and-analysis/studies/TOCTA_EAP_web.pdf.
8   United Nations Environment Programme, "Organized Crime Trade Worth over US$30 Billion Responsible for up to 90% of Tropical Deforestation," United Nations Environment Programme, Sep. 27, 2012, http://www.UNEP.org/documents.multilingual/default.ASP?DocumentID=2694&articleID=9286&l=zh.
9   Global Witness, "Peru's Deadly Environment," Global Witness, Nov. 2014, http://www.globalwitness.org/perudeadlyenvironment/.
10 United Nations Environment Agency and INTERPOL, "The Environmental Crime Crisis," United Nations Environment Agency and INTERPOL, 2014. For more information on the Liberian case, see the Global Witness report to the United Nations Security Council in June 2005: "Timber, Taylor, Soldier, Spy," https://globalwitness.org/sites/default/files/import/TimberTaylorSoldierSpy pdf.

1.3   Undermining Efforts to Address Climate Change

Natural forests play a vital role in combating climate change. The world's forests absorb about half of the total carbon emissions from fossil fuels, and tropical forests alone absorb more than a billion tons of carbon from the atmosphere each year.[11] However, previous studies estimated that deforestation accounted for 20% of global artificial emissions of carbon dioxide.[12] According to a joint report released by the United Nations Environment Programme and Interpol in 2012, reducing deforestation, especially illegal logging, is the fastest, most effective and least controversial way to reduce global greenhouse gas emissions.[13]

1.4   Damaging Biodiversity

In addition to causing significant increases in greenhouse gas emissions, deforestation threatens biodiversity, especially in the tropics where the forest species account for nearly half of the world's creatures.[14] For example, in Malaysia, the lax regulation and poor law enforcement have allowed logging companies systematic violations of Sarawak forestry laws, by logging at a rate twice or even three times higher than that of sustainably managed logging.[15] It is estimated that only 5% of Sarawak's virgin forests have not been destroyed,[16] in sharp contrast to the Sarawak government's claim that the region has an 84% forest cover.[17] The region is significant for biodiversity, inhabited by endangered

---

11   Pan and Yetal, "A Large and Persistent Carbon Sink in the World's Forests," *Science* (2011).
12   Chatham House, "Illegal Logging and Related Trade: Indicators of the Global Response," Chatham House, 2010, http://www.chathamhouse.org/sites/files/chathamhouse/public/research/energy.20%environment20%and20%development/0710bpp_illegallogging.pdf.
13   United Nations Environment Agency and INTERPOL, "Green Carbon, Black Trade: Illegal Logging, Tax Fraud and Laundering in the World's Tropical Forests: A Rapid Response Assessment," United Nations Environment Agency and INTERPOL, 2012, p.13.
14   For example, see Lindsey, R, "Tropical Deforestation," NASA, March 30, 2007, http://earthobservatory.nasa.gov/Features/Deforestation/.
15   Global Witness, "An Industry Unchecked: Japan's Extensive Business with Companies Involved in Illegal and Destructive Logging in the Last Rainforests of Malaysia," Global Witness, September, 2013.
16   Global Witness, "An Industry Unchecked: Japan's Extensive Business with Companies Involved in Illegal and Destructive Logging in the Last Rainforests of Malaysia," Global Witness, September, 2013.
17   Sarawak Chief Minister, "Forestry in Sarawak," Sarawak Chief Minister, http://chiefministertaib.sarawak.gov.my/en/perspectives/the-environment.

species including orangutans, elephants and rhinoceros.[18] Reports published in the past one or two years have also pointed out that a number of species, such as Siamese rosewood in the Mekong region,[19] Siberian tigers and Amur tigers in Russia,[20] and golden monkeys in Myanmar,[21] are threatened by illegal logging to varying degrees to lose their habitats.

3   China is Closely Connected to Global Timber Trade

With the growth of demand for timber, China plays an important role in global timber trade. This entails the duty to ensure that such trade is legal rather than illegal. China should take action to deal with illegal logging and related trade for reasons including but not limited to those mentioned below. Other reasons include improving relations with timber producing countries, demonstrating China's leadership in international issues such as anti-corruption and response to climate change, etc.

3.1   *China is the World's Largest Importer of Forest Products, Which Involve a Large Number of Illegal Timber*

China is currently the world's largest importer of wood and forest products.[22] Take tropical wood for example, China imports 60% of the world's tropical logs and a third of its tropical timber products,[23] a lot of which are illegal imports due to the lack of effective legal restriction. According to research, more than half of China's current supply of log materials come from high-risk countries

---

18   WWF Global, "Wildlife in Borneo," WWF Global, http://wwf.panda.org/what_we_do/where_we_work/borneo_forests/borneo_animals.

19   Environmental Investigation Agency, "Routes of Extinction: The Corruption and Violence Destroying Siamese Rosewood in the Mekong," Environmental Investigation Agency, May, 2014.

20   Environmental Investigation Agency, "Liquidating the Forests: Hardwood Flooring, Organised Crime, and the World's Last Siberian Tigers", Environmental Investigation Agency, 2013.

21   Fauna & Flora International, "First Video Footage of Myanmar Snub-nosed Monkey Captured," Fauna & Flora International, March, 2014, http://www.fauna-flora.org/news/first-video-footage-of-myanmar-snub-nosed-monkey-captured/.

22   Environmental Investigation Agency, "Appetite for Destruction: China's Trade in Illegal Timber," Environmental Investigation Agency, 2012, p.4.

23   Global Witness bases its estimates on data from tropical timber producers, UN COMTRADE Database and China's import data. Tropical timber products include log, sawn timber, veneer, plywood and mould board.

where illegal logging control and forest governance are weak.[24] According to conservative estimates, China imported at least 18.5 million cubic meters of illegal logs and sawn timber worth US$3.7 billion in 2011, accounting for 10% of its total imports of wood products in 2011.[25] The real figures are much larger, because the above figures only include China's imports of logs and sawn timber, which account for only 45% of China's total imports of wood products.[26]

A new report released by Global Witness in February 2015 says China's craze for mahogany furniture has spawned smuggling of Cambodian timber worth millions of dollars.[27] The report reveals Cambodian tycoon Okhna Try Pheap's illegal logging network in collusion with government and law enforcement officers, illegally cutting down precious wood such as Siamese rosewood. Extensive logging has accelerated the extinction of tree species, and forced aboriginal people and forest communities to move away from the resources they made a living with. According to statistics, Cambodia exports 85% of its timber to China, with mahogany exports up 150% between 2013 and 2014.

China is also a major exporter of forest products, with the United States, the European Union and Japan as its main customers.[28] However, domestic consumption still accounts for the largest share despite large amounts of wood exported after processing. According to the data of the State Forestry Administration of China, export accounted for only 17.56% of China's total consumption of timber products in 2013, while the rest was domestic consumption.[29] In view of the current international economic situation and the rapid development of its domestic market, China's domestic consumption of forest products will continue to rise. Given China's huge market of import and domestic consumption, China's has a crucial influence on timber trade. Whether this will direct timber trade to sustainable development depends on the choice of the Chinese government and consumers.

---

24   Environmental Investigation Agency, "Appetite for Destruction: China's Trade in Illegal Timber," Environmental Investigation Agency, 2012, p.5.
25   Environmental Investigation Agency, "Appetite for Destruction: China's Trade in Illegal Timber," Environmental Investigation Agency, 2012, p.7.
26   Environmental Investigation Agency, "Appetite for Destruction: China's Trade in Illegal Timber," Environmental Investigation Agency, 2012, p.7.
27   Global Witness, "Cost of Luxury," Global Witness, http://www.globalwitness.org/costofluxury/.
28   Environmental Investigation Agency, "Appetite for Destruction: China's Trade in Illegal Timber," Environmental Investigation Agency, 2012, p.5.
29   State Forestry Administration, *China Forestry Development Report 2014* (Beijing: China Forestry Press, 2014), 136.

The United States and the European Union are also major consumers of forest products, many of which are processed and exported from China. In order to deal with illegal logging and trade, the United States and the European Union have taken some positive measures to ban trade of illegally logged timber, and achieved certain results. However, in the management of international timber trade, without the participation of China as a major importer, processor and consumer, the efforts made by timber producing countries and other consuming countries to deal with illegal logging and trade will only have limited effect.

3.2 Recognition and Respect for Relevant Laws of Timber Producing Countries

China has always highly recognized and respected the laws of other countries. Several policies concerning overseas investment and trade[30] have clearly stated that Chinese enterprises should comply with the laws and regulations of the target country (region) in their overseas investment and trade. Therefore, China should incorporate its international timber trade policy into this principle by refusing timber that violates this principle to enter China. By doing so, China can fulfill its commitment to help timber-producing countries maintain laws and protect natural resources, which will help strengthen relations between China and timber producers.

3.3 Playing a Leading Role in International Action to Combat Corruption and Climate Change

Since 2014, China has become increasingly willing and motivated to participate in international anti-corruption efforts. On November 8, 2014, the 26 APEC Ministerial Conference adopted the Beijing Declaration on Fighting Corruption, the first international anti-corruption declaration drafted by China. China is a member of the Group of 20 (G20) that approved the Anti-Corruption Action Plan 2015–2016 during its 9th summit on November 16, a milestone in international cooperation against corruption. It is worth noting that the G20 Anti-Corruption Action Plan 2015–2016 clearly states that primary forestry is a high-risk sector vulnerable to corruption and that the

---

30   Such as: Interim Measures for the Supervision and Administration of Overseas Investment of Central Enterprises, Article 5: "(Overseas investment) shall abide by the laws and policies of the host countries (or regions)"; Guidelines for Environmental Protection in Foreign Investment and Cooperation, Article 5: "Enterprises shall understand and observe provisions of laws and regulations of the host country concerning environmental protection"; Green Credit Guidelines, Article 21: "... make sure project sponsors abide by applicable laws and regulations on environmental protection, land, health, safety, etc. of the country or jurisdiction where the project is located"; etc.

G20 members should pay special attention to it in 2015/2016 and work with other countries to reduce corruption in primary forestry in order to ensure a sustainable supply of forest resources.[31] Against this background, China may choose forestry as one of the starting points in its participation in the international anti-corruption campaign, by supporting timber producing countries in their efforts to combat corruption concerning forest industry through legislation prohibiting illegal timber imports, and assisting governments of timber producing countries in enhancing their governance and law enforcement capacity. It will show China's determination to participate in international anti-corruption campaign, its soft power in international affairs and the image of a big responsible country.

China has also played a leading role in tackling climate change. During the APEC meeting in 2014, China and the United States, the world's top 2 emitters of greenhouse gases, issued China-US Joint Statement on Climate Change, setting out their respective objectives on climate change beyond 2020. At the G20 summit in 2014, China, the United States and other major countries put climate change on the agenda, confronting climate issues and making commitments to tackle them together. These efforts show that China is dealing with climate change with positive attitude and firm determination. As noted earlier, illegal logging and trade have hindered the international community's efforts to combat climate change, whereas combating illegal logging and its trade will make a positive contribution to the fight against climate change. And China can do much to crack down on illegal logging and its trade.

4   Efforts at Home and Abroad

Since the Group of eight (G8) summit in 1998 first put forward illegal logging as an important international issue, illegal logging and related trade have attracted more and more attention by all parties. China and other timber producers and consumers have taken steps to deal with the problem.

4.1   *International Efforts*
The action taken by timber producing countries against illegal logging in their countries is very important. At the same time, the world's major timber importers and consumers are taking action to deal with the problem. For one thing,

---

31   G20 Anti-Corruption Working Group, "G20 Anti-Corruption Action Plan 2015–2016," G20 Anti-Corruption Working Group, http://www.g20australia.org/official_resources/2015_16_g20_anti_corruption_implementation_plan.

some importing and consuming countries help timber producing countries improve their forest governance and forestry enforcement capabilities by providing training and financial support, such as The European Union Forest Law Enforcement, Governance and Trade Action Plan (or EU FLEGT Action Plan) launched in 2003 and the voluntary partnership agreements between the EU and its partners (FLEGT-VPA). For another, these countries and regions take measures to ensure that their procurement and investment practices do not inadvertently contribute to this problem. These measures include national or regional legislation explicitly banning illegal timber imports and trade. Among them, Amendment to the US Lacey Act entered into force in May 2008, which stipulates that it is unlawful to import, export, transport, sell, receive, acquire, or purchase any plant taken, possessed, transported, or sold in violation of any law, treaty, or regulation of the United States or relevant countries; The European Union Timber Regulation was adopted in December 2010 and came into force in March 2013, prohibiting the sale of illegally logged timber and its manufactured forest products and products processed from such timber in the EU market; Australia Illegal Logging Prohibition Act of 2012 advocates prohibiting the import or processing of illegally logged timber.

4.2   *Domestic Attempts*

China has taken action to deal with illegal logging and related trade issues, including *A Guide on Sustainable Overseas Silviculture by Chinese Enterprises jointly issued by the National Forestry Administration and the Ministry of Commerce* in 2007, *A Guide on Sustainable Management and Utilization of Overseas Forest by Chinese Enterprises* issued in 2009, and *A Guide on Sustainable Overseas Trade and Investment of Forest Products by Chinese Enterprises* which is being drawn up. One of the main focuses of the third guide, still at the draft stage, is "trade," which can be seen to some extent as a rare statement by the Chinese government that it would take measures to deal with illegal logging and related trade. The drafting and consultation processes have been more open than in the past in terms of transparency and consultation with different stakeholders.

The guides are not legally binding and are implemented voluntarily. The two previously published guides, while lauded, have limited restrictions on Chinese enterprises in practice and have failed to play a significant role in changing enterprises' conduct. However, due to the lack of domestic regulations and standards on illegal timber trade, if the new guide can provide a useful supplement in this regard, it will lay a good foundation for further work. Nevertheless, as far as the current draft is concerned, *A Guide on Sustainable Overseas Trade and Investment of Forest Products by Chinese Enterprises* seems to have failed to provide a real solution to the principal problem in the formulation of some provisions. The most crucial point is that, while the guide requires Chinese enterprises to "refrain from purchasing timber and its products of illicit sources," its application is limited to "Chinese enterprises involved in trade, investment and other activities abroad in forest products." This scope of application is similar to the provisions of the previous two guides, which do not include Chinese domestic enterprises importing or processing timber from abroad. In fact, the key point in dealing with illegal logging and related trade is not how much illegal timber Chinese enterprises are buying abroad, but how much illegal timber it is absorbing from all over the world. Therefore, the new version of the guide will be more likely to have a positive impact by changing the scope of application of the guide to "applicable to Chinese enterprises involved in trade, investment and other activities abroad in forest products as well as Chinese enterprises that purchase wood products from outside the country," and explicitly requiring that "all Chinese enterprises should not purchase illegal timber and its products, and that strict due diligence be carried out, to ensure that their supply chains are free of illegal forest products."[32]

In addition to issuing the guides, the Chinese government has reached bilateral agreements with a number of timber producing countries, including the Memorandum of Understanding (MOU) with Indonesia in 2002, the MOU with Myanmar in 2006,[33] and the bilateral mutual recognition mechanism currently under discussion with Papua New Guinea, and so on. The development of the bilateral cooperation shows that China has realized the existence of illegal logging and related trade in some countries, and is willing to cooperate with producing countries. However, illegal logging and related trade are serious and urgent global problems in which China needs to play a leading role in the world with mandatory global response policy.

5 Suggestions

To sum up, dealing with illegal logging and related trade is not only extremely important for the world, but also beneficial for China in many aspects: it will enhance China's relations with timber producing countries, improve its

---

32 Global Witness, "Submission to the Chinese State Forest Administration Regarding the Forthcoming Timber Trade and Investment Guideline," Global Witness, January 2014, https://www.globalwitness.org/zh-cn/archive/submission-chinese-state-forest-administration-regarding-forthcoming-timber-trade-and/.

33 Environmental Investigation Agency, "Appetite for Destruction: China's Trade in Illegal Timber," Environmental Investigation Agency, 2012, p.3.

international reputation, and build up the positive image of a responsible big country; it will reduce the environmental crime rate in the Asia-Pacific region; it will show China's positive action in international cooperation on issues such as anti-corruption and climate change; and so on.

China has made useful attempts on the issue, but they are limited in scope and effectiveness. As a major player in timber trade, China needs to take active steps on the fundamental problem of illegal logging and related trade: cutting off the channel of timber from illegal sources. Since illegal logging and trade are serious global problems, these actions should be global and legally binding.

There is a lot that China can do next, including involving the forestry administration and relevant authorities, such as the Ministry of Commerce and the Customs, in the policy consultation and joint operations to combat illegal timber trade, promoting the information disclosure on the timber supply chain of Chinese enterprises engaged in global timber trade; opening up to more public scrutiny, etc. One of the most important actions is to promote legislation to ensure that China's trade and investment activities do not inadvertently contribute to the growth of illegal logging and related trade. The law should contain two key elements: one is to state clearly that China refuses to import illegally logged timber; the other is to require Chinese enterprises to conduct strict due diligence on the supply chain of their wood products to ensure they are free of illegal forest products and to release results of the due diligence.

As mentioned above, there are already many experiences in the world in prohibiting illegal timber import through legislation. China can draw up laws suitable for its conditions by learning from existing experiences. If China clearly states its rejection of illegal timber in the law, it will show China's determination to crack down on illegal timber trade, thus winning more international recognition and support for China. More importantly, it will provide legal basis for law enforcement agencies. For example, customs authorities will have the power to deny illegal timber entry.

Provisions in the law that require Chinese enterprises to carry out due diligence on the timber supply chain can disperse and move forward the pressure of law enforcement, so that Chinese enterprises can ensure that they are engaged in responsible timber trade by establishing the due diligence system and conducting it when participating in international timber trade. Due diligence in wood supply chain refers to the process of establishing timber supply chain and collecting investment information on the initiative of enterprises investing or purchasing timber abroad, carrying out risk assessment, and taking measures against identified risks, also including information disclosure on the process and results of due diligence. The concept of corporate due diligence has a long history in global commerce. For example, international financial and anti-money-laundering standards require financial institutions to conduct due diligence on the identity of their clients to ensure that their money does not come from illegal and criminal activities; enterprises which purchase minerals from areas with high risk of conflict should conduct due diligence on their supply chain, etc. There is also a lot of international experience on how to carry out due diligence on wood supply chain.

The legislation on illegal timber import is an international trend. It is also the positive response of China, the world's largest importer of forest products, to the serious situation of illegal logging, which will bring many benefits for China. Therefore, we suggest that the Chinese government put the legislation on its agenda as soon as possible.

CHAPTER 15

Chinese Involvement in Brazil's Development: Massive Investments Bring Environmental and Cultural Challenges

*ZHOU Lei and Petras Shelton ZUMPANO**

Abstract

Based on field investigations, interviews and document research, this chapter analyzes the difficulties and challenges that China has faced or will face when investing in agriculture, infrastructure and energy projects, as well as the underlying ethnic politics, ecological protection, international organizations, elite knowledge production and media communication. Based on the analysis, this study reviews the economic interactions between China and Brazil and the overall responses of the Brazilian government and civil society.

Keywords

China – Brazil – overseas investment – environmental impact – indigenous challenge – corporate diplomacy

In 2012, as part of the project "Environmental Impact and Cultural Adjustment of China's Overseas Investments" of the independent think tank Oriental Danology Institute, the author cooperated with the colleagues of University of Brasilia and the BRICS think tanks in studying several Brazilian states with intensive indigenous resources in order to reveal the environmental pressures and cultural challenges underlying China's overseas investments. This study

---

* Zhou Lei, Doctor of anthropology and founder of the independent think tank Oriental Danology Institute, has designed this research, conducted the field investigation and written this chapter. Petras Shelton Zumpano, strategic adviser, researcher of the Rio Program of UN-HABITAT (United Nations Human Settlements Programme) and doctoral candidate of Xiamen University, has contributed to this research by contacting local institutions and governments for the two field investigations and analyzing the views of Brazilian media and scholars.

is based on the interviews with the representatives of the Brazilian government, think tanks, congress and NGOs as well as the representatives of Chinese companies and government in Brazil. We have also visited and investigated the indigenous regions. In June 2013, the author visited the Sao Paulo, Brasilia and Mato Grosso States of Brazil and interviewed relevant representatives so as to explore how foreign investments influenced the ecologically sensitive zones of Brazil and how local people dealt with the external ecological risks. In 2014, my colleague Petras Shelton Zumpano continued the investigation in Rio de Janeiro and Sao Paulo to collect and analyze the responses of the local media, government and civil society to the state visit of the Chinese President Xi Jinping.

During the 2013 field investigation, the author travelled from Brasilia to Mato Grosso with the local Xavante driver and guide, who worked in the Brazilian government. According to him, the name "Xavante" was created by the Portuguese colonists, while the locals call themselves "A'uwe Uptabi," which means the "real people." Due to the persecution of the colonists, the A'uwe Uptabi people have migrated and settled in Brasilia and other regions, where China has invested in soybean plantations. In the adjacent Bahia State, China has made even larger-scale investments in agriculture. Towards the end of the investigation, my Xavante guide told me in a letter that he would join the local socialist party and run for the local decision-maker representing the aborigines. Throughout 2014, he often shared photos of making public speeches to different electoral districts and communities on Facebook, through which I learned more about the electoral campaigns and developments in Brazil's jungles, indigenous communities, agricultural regions and the capital Brasilia.

This observation reflects the cultural background of China's investment in Brazil, i.e. a Brazil featuring increasing globalization, political pluralization, interest differentiation and social differentiation. Here, no investment would have a static and closed virgin land, but an increasingly dynamic and global scenario. Although China's investment in Brazil has shown massive growths in recent years,[1] the key to a long-term success for China's capital is learning how to speak, act and think in the new investment scenario. In this sense, Chinese

---

1 According to the data released by SECEX, from January to August 2014, Brazil's export to China amounted to a total value of 31.7 billion USD, in which 83.7% was contributed by soybeans, meat products, iron ores and petroleum. In the first half of 2014, 89% of the soybeans imported by China came from Brazil, showing a great increase from the 50% in 2013. In 2012, Brazil exported half of its iron ore and soybean output and one quarter of its crude oil output to China. The soybean export had an increase of 40% in the first three quarters of 2013. In 2013 and 2014, China was estimated to produce 12–12.5 million tons of soybeans. In 2009, China replaced the United States as the biggest trade partner of Brazil. In the first three

investors have become "corporate diplomats" to a certain degree. In July 2014, President Xi Jinping proposed a grand strategic design on the BRICS Summit to build a transcontinental railroad to connect the Atlantic and Pacific coasts of South America. According to the bulletin of the Brazilian federal government, in October 2014, China Railway Eryuan Engineering Group bade for two railroads, i.e. Railroad EF-354 connecting Sapezal and Porto Velho and Railroad EF-170 connecting Sinop of the Moto Grosso State and Itaituba of the Bahia State, which will be used to connect the Atlantic and Pacific oceans and build a transcontinental logistics network. This kind of super projects and massive investments will influence the local political pattern in a significant way.

1 China, Brazil and Latin America: Ecological Worries Underlying Massive Investments

On the 6th BRICS Summit, the five countries decided to jointly build a development bank with a reserve fund of 100 billion USD. On the conference of the Community of Latin American and Caribbean States (CELAC), China announced an additional investment of 35 billion USD in Brasilia, including 20 billion USD to be invested in infrastructure facilities. The cooperation between China and Brazil as well as other Latin American countries has always been a consensual one, but how long can this cooperation last?

In November 2014, China and Peru signed the agreement to build the transcontinental railroad with a distance of 5,300 kilometers. The railroad will bypass the Panama Canal and enable the direct trade transportation between Atlantic and Pacific. The railroad will greatly change the international trade and global geopolitical pattern but will also confront massive environmental pressures and social risks, since the construction will affect vast tropical rainforests and biodiversity regions. In this super project, 2,900 kilometers are located in Brazil, for which China will invest at least 10 billion USD while Brazil will only need to invest 2.3 billion USD to build an 880-km railway. Meanwhile, China has proposed to build a Nicaragua Canal with an investment of 50 billion USD and a distance of 275 kilometers. As China and Latin America deepen their economic integration, China-Brazil interaction will be influenced by the overall economic integration between China and Latin America. From 2010 to 2013, China steadily increased its investments in Latin American countries, with at least 6 billion USD invested into local infrastructure construction of

---

quarters of 2013, Brazil's export to China surpassed the total export of Brazil to the European Union.

Peru. In October 2014, CELAC issued a report proposing that to sustain the economic development from 2012 to 2020, Latin American countries should invest at least 6.2% of their GDP into infrastructure construction, amounting to approximately 320 billion USD. At present, the infrastructure construction in Latin America including Brazil is mainly intended to supply ores, energy resources, agricultural products and other bulk commodities for China, especially gold, copper ores, tin ores and petroleum in Peru and soy beans and iron ores in Brazil. Besides, China has purchased 40% of Chile's copper mines and lent Venezuela a loan of 50 billion USD to obtain local petroleum resources. The way China invests and obtains resources in Latin America has created a new economic growth mode, i.e. "premature deindustrialization" as defined by the Harvard development specialist Dani Rodrik.[2]

As the Brazilian government estimates, once the transcontinental railroad is built, approximately 30 USD will be saved for every ton of commodities transported from Brazil to China. However, this estimate does not take into account the consequences of infrastructure construction, such as habitat degradation, biodiversity loss and the ecological and cultural impacts of local economic transformation. During the construction of the transcontinental railroad, the 2,575-km railroad passing through the tropical rainforests and the Andes ecoregions from Sao Paulo to Lima will greatly change the rainforest distribution and lead to significant habitat fragmentation, soil degradation, species extinction and damages to indigenous livelihood. As many designs are intended for mining development, illegal mining and resource exploitation are likely to be stimulated among the locals. This trend will be enhanced by drug trafficking, revolts and local ethnic politics in Latin American countries. Consequently, a chaos will be created in the environment, society, economy and politics of Latin America. In the 1970s, Brazil constructed the 4,000-km highway crossing Amazon rainforests, which caused irreversible ecological damages to the primitive tropical rainforests. According to the estimates of Mongabay, a database for rainforest coverage and environmental impact, approximately 2 million hectares of forests disappear every year because of various construction projects in Latin American countries, since the small countries along the Andes Range have the densest distribution of rainforests.[3]

When China started its investment in Brazil and other Latin American countries, China did not conduct any research or make any long-term plan

---

2 Eduardo Porter, "Slowdown in China Bruises Economy in Latin America," *New York Times*, December 16, 2014.
3 Kamilia Lahrichi, "China's Asphalt Footprint Threatens Amazon Biodiversity," *Asia Sentinel*, February 4, 2015.

concerning Latin American history and culture, ethnic politics and environment as well as the industrialization of ecological societies and the modernization of traditional oriental civilizations. Instead, China adopted the western model featuring "primary industrialization," pollutant export and energy agency, which is a non-sustainable model basing the "sustainable" domestic development on the non-sustainable overseas development. To break the deadlock, China should start by studying the overseas governance scenario to explore the soft environment of investment, i.e. such non-economic elements as ecological dilemma, land policies, ethnic politics, cultural syndromes and ethnographic elements. Based on the understanding of these elements, China can then reconsider its project designs.

2 Political-Cultural Elements and Resource Traps Underlying China's Investment in Brazil

In August 2014, my colleague of this research Petras Shelton Zumpano analyzed the responses of Brazilian intellectuals to the state visit of the Chinese President Xi Jinping when attending the BRICS Summit. A typical view was put forward by Adrian Hearn, a renowned expert in Latin American studies. Hearn pointed out that the complicated reliance model between China and Brazil still has severe structural problems, which will hinder the interactions between China and Brazil. Around 2000, with the growth of its national power, Brazil showed a stronger desire to participate in international issues and play a bigger role in such fields as trade, international finance, climate change and nuclear weapon proliferation. China, however, has always maintained a relatively neutral and conservative position, dealt with the "China Threat" claim of the international society in a low-key and moderate manner and defined its own development as "rising," i.e. a status between developed countries and developing countries. For a long time, China and Brazil have adopted different approaches: Brazil considers its relationship with China as a strategic multilateral one, while China defines its relationship with Brazil as an important bi-lateral one. As a result, the economic and trade interactions between China and Brazil are characterized by one-sided enthusiasm, with China playing a more active and driving role and Brazil making passive responses and sometimes showing hesitations and reluctances. The one-sided and partial strategic partnership between China and Brazil has caused various problems in such aspects as mechanism cooperation, social cooperation, business and trade, investment interaction, military cooperation and diplomatic collaboration.

During the investigations, it was found that the political background described above could have direct impacts on China's investments. A good example is the 2 billion USD investment of Chongqing Grain Group in the Bahia State. As of the end of 2014, China had made no investment in local plantations. The Chinese company originally planned to grow and process soybeans and build infrastructure and storage facilities. Up to now, however, the local government has only allocated 100 hectares of land as the factory site, with little progress in other parts of the plan. The Bahia State claimed that China's investment should pass municipal examination and approval and meet strict requirements for environmental protection. Therefore, the funds in the investment have not been credited. Chinese investors are confronted with such obstacles as bureaucracy, corruption, low efficiency, economic slowdown and the barriers in political mutual trust between China and Brazil. As a matter of fact, Brazil has taken precautionary measures against China's bulk purchases of real estates, mines and land. When the author interviewed the members of the Brazilian congress, they mentioned a number of regulations restricting foreigners purchasing land, especially the Chinese.

The aborigines, local governments and foreign investors have different understandings of resources, land and livelihood. For the aborigines, land is part of their possessions, consisting a world together with their culture, tradition, faith, art and language. Many indigenous communities and organization leaders call the outsiders "invaders." The indigenous rights organizations often quote the Palestinian poet Mahmoud Darwish to show their determination to resist aggression. As for the governments, they promised to give the aborigines equal rights in the Constitution but have failed to completely fulfill the promise. The 1988 Constitution has described the indigenous territories and confirmed the indigenous rights and interests. In the actual political games, however, it is often difficult to mark and confirm indigenous territories and to make and enforce relevant laws. Attracted by massive foreign capital, the governments often issue licenses to external investors without prior negotiation with stakeholders, which have caused disputes and conflicts. During the investigations in Sao Paulo and Brasilia, the author witnessed frequent protests of ethnic minorities. In Brazil, different political views spread quickly through the media, on the streets and in the villages. For Chinese investors, Brazil is a land of rich mineral reserves, agricultural products and energy resources rather than a habitat of the aborigines, a land of miracles and a Noah's Ark of species. All regions are developed as a project and the only relationship between the investors and the aborigines is the economic compensation. China has paid a heavy price for ignoring the long colonization history and the complicated post-colonization cultural pattern in Brazil. In the investigations, it was found that the indigenous villages showed either indifference or distrust towards the massive investments and projects, which is defined as one of the symptoms of the environmental protection deadlock by Dr. Henyo, the director of the International Institute of Education. The Brazilian Constitution has an article for indigenous rights and interests and the local governments have tried to build a pan-Amazon ecological protection network and specify the aborigines' rights in managing their habitats and the environment. However, all these efforts have become invalid due to the overwhelming external capital. To control the environmental impact of foreign capital is a long-term project. Foreign investors and domestic interest groups in Brazil use various policies and incentives to disintegrate the resistance alliances of the civil society. Adriana, the executive secretary of the Society and Environment Research Institute of Brazil, pointed out that many projects with catastrophic ecological impacts have the same operating logic. First, the investors would obtain the license issued by the governments. Due to the blurry power boundary between the federal government and local governments, many projects were approved without the joint negotiation of the federal government, local governments, grassroots organizations, NGOs and local residents. Then, many international organizations would try to persuade the public by explaining why the projects are necessary, how the local communities can benefit from the projects and why the aborigines should learn to adapt to the modern society rather than discuss whether the investments are legal or not and how the investments and developments should try to minimize their ecological impacts. According to Adriana, the global economic crisis has made environmental protection campaigns more difficult, because governments now give the priority to attracting investment and stimulating economy. Moreover, there are fewer environmental pressures from the public. As a result, foreign companies care little about the ecological commitments and the letters of commitment have little binding force.

Based on the investigations and interviews, we have reasons to believe that the potential ecological risks in the Pan-Amazon zone of South America will be closely related to massive capital investment and urban expansion. The environmental protection forces of the globe cannot rival the international capitals in terms of completeness, efficiency and ambition. Nowadays, it is easy to launch large-scale projects but extremely difficult to replicate miniature environmental protection cases. As a result, many regions with ecological values have become "ecological isolates." If Chinese investors fail to understand the ecological, political, ethnic and international scenarios underlying the resources they attempt to develop, they are doomed to fall into the resource traps of overseas investments.

3 Resource Politics in Brazil: the Power Chaos of the Indigenous Policies

From 2001 to 2014, the economic interactions between China and Brazil mainly focused on soybeans, petroleum, ores and other resources. Brazil is eager to stop this low-end export model and make more exports to China in such high-end fields as infrastructure, airplanes and high and new technology. China, however, also aims to export in these fields. The last decade witnessed massive growths in China's investment in Brazil. For example, CNOOC (China National Offshore Oil Corporation) and CNPC (China National Petroleum Corporation) invested in exploring the offshore oil resources of Brazil and CCB (China Construction Bank) purchased 72% of the shares of local Brazilian banks with an investment of 726 million USD. China's investment now covers more fields and takes more approaches, but the main purpose is still acquiring resources. Margaret, the director of the US-based think tank Intercontinental Dialogue pointed out that in terms of overseas investment, the Chinese government shows an image that the Chinese government and the state-owned capital attempt to monopoly and acquire bulk resources. Frequent international mergers and acquisitions of China have created a local panic to some degree. As a result, China has to try some unconventional methods to obtain more shares or share options so as to save costs or skip some procedures.

Against this background, there is a more urgent need for China to understand the indigenous policies of Brazil. Article 231 of the 1988 Constitution of Brazil, designed for the Native Americans, stipulates that the land inhabited by the aborigines belongs to the federal government but the aborigines own the properties on the land. After negotiating with the aborigines, the government and the congress can jointly make development policies for hydroelectricity or mining investment.[4] In 2002, Brazil modified the Convention 169 of the International Labour Organization to further specify indigenous affairs and local negotiations. Currently, approximately 13% of the Brazilian territory involves indigenous rights and interests, among which 97% is located in the Amazon, especially the Northern Amazon. With the popularity of indigenous rights movements, over 20% of the territory is likely to be involved in the future. These regions, however, are sparsely populated. Some indigenous rights groups have claimed that the Native Americans in Guarani may create an independent sovereign country. There are fierce struggles over land ownership and

---

4 In June 2013, we interviewed Padre Ton, a member of the House of Representatives of the Brazilian Congress, for over an hour. Also present were the representatives of indigenous women's rights and Xavante rights. We are grateful to Padre Ton, who contributed significantly to the description of indigenous policies of this study.

land use in Brazil. The ethnic minority groups aim to dominate more regions, while the government also tries to expand their dominion so that they can make more commercial development and infrastructure construction.

The current land policies and land ownership conflicts in Brazil make indigenous land management and protection difficult. The chemicals and pesticides used in modern agriculture and the development of farming and animal husbandry have caused the decrease in vegetated land. For example, the Rondônia State of the Amazon is now raising over 12 million heads of cattle. To sustain this industry, the state has cleared large areas of forests. Another threat to the Amazon is sugarcane farming, which provides raw materials for generating new energy. However, the Pará and Flexa Riberiro States have submitted a proposal to the congress for clearing more forests and developing sugarcane farming in transitional zones of the Amazon rainforests. In 2013, the National Indian Foundation of Brazil submitted a new map of the land owned by the indigenous peoples. Similar maps have also been proposed by the National Institute of Colonisation and Agrarian Reform of Brazil and other organizations. Among these land ownership struggles, some indigenous groups have adopted more strategic protest policies. The Munduruku people, for example, chose to protest in the construction site of the Belo Monte Hydropower Station of the Pará State so as to obtain more external support.

4 The Land of Wrath: Growing Opposition to Primary Industrialization

In April 2012, the landless peasants in Bahia started a revolt. They attacked the city hall and occupied large areas of "ownerless" land. The land acquired by Chongqing Grain Group with 1.6 billion USD to plant and process soybeans was just located in this region. In Brazil, political protests are often inflamed by a small incident. These protests involve complicated factors such as indigenous rights, poor governance, external capital, social disorder and social differentiation. Chinese investors can also get involved in some protests. Therefore, China should develop a mechanism of balancing interests and making multi-side dialogues so that its overseas investments can avoid falling into the trap of joint protests. The indigenous peoples in Brazil have a well-established organization, a diverse system of discussion and a direct way of expressing their concerns. There are over 220 indigenous groups, which speak over 180 languages. Despite the wide distribution, they have established some specialized organizations, e.g. the Coordinator of Indigenous Organizations of the Amazon River Basin (COICA), the Andean Coordinator of Indigenous

Organizations (CAOI), Central American Indian Affairs Office, the Indigenous Rights Expressing Organization, the Intellectual Property Rights Program of the Indigenous of Brazil's Indigenous Research Institute, and the Inter-Ethnic Memory and Native Science Association. These indigenous rights organizations and intellectual empowering organizations play a significant role in unifying the aborigines and spreading information. They can also represent the ethnic groups and communities to express their concerns to external investors, local governments and the international society and to negotiate and solve problems concerning the environment, development and cultural protection. The indigenous groups are also urging the governments to delimit, map and confirm the indigenous territories so as to better protect their ethnic interests and rights.

Given the interest conflicts, information gaps and divided opinions of the aborigines, rights protection groups, government officials and foreign investors, environmental protection in Brazil is faced with great difficulties. Adriana, the director of the Alliance for a Sustainable Amazon, pointed out that the laws of Brazil are not effectively enforced, leaving loopholes for violations and noncompliance. Currently, the political power groups of Brazil have divided opinions on how to delimit the indigenous territories. Theoretically, it is getting more difficult and costlier to make illegal investments on the marked indigenous territories. In the capital, the indigenous groups often protest in front of different departments of the federal government. However, many large-scale investment projects are still under construction, such as the Belo Monte Hydroelectricity Station, the biggest gold mine in Brazil invested by Canada and the 200,000-hectare soybean plantation with an investment of 5.75 billion RMB by Chongqing Grain Group.

5 Conclusion: Chinese Investors Need to Understand Local Concerns and Ecological Adaptation

There has been a great increase in China's investment into Brazil's agriculture, which has created both opportunities and risks. The Brazilian policy makers define China as a rapidly rising power and economy and relate China's investment directly to the domestic conflicts, social affairs and environmental protection in Brazil. This misconception of China may hinder China's economic and political cooperation with Brazil and other emerging markets.

Generally speaking, China has a more positive image in Brazil than in other developed countries. However, a 2013 survey by the Pew Center shows that among 960 participants from different regions of Brazil, nearly 50%

considered China as a partner, 10% saw China as an enemy and 51% did not like China's investment. In the field of international investment and politics, reputation and credibility are important elements of power and can greatly affect a country's position and interest in the international society. The intellectuals and media of Brazil are concerned about such issues as "de-industrialization" and "interaction levels" in China-Brazil economic interactions. They are worried that when making investments, China may only consider its domestic economic growth, industrial innovation, export and resource acquisition but may harm Brazil's independence, self-reliance and innovation in the economic globalization. Another survey conducted by Gallup in 2012 for the BRICS countries included a question whether environmental protection should be the first priority as a factor influencing economic growth. To this question, 57% of Chinese participants answered yes while 83% of Brazilian participants answered yes. Since 2008, several protests have broken out against foreign investors who were building soybean plantations in Brazil. The protestants blocked highways with trucks and claimed that agricultural development destroyed the environment and traditional livelihood of the locals. Although these protests did not involve Chinese investors, China is very likely to be the next target if it does not shoulder its environmental and social responsibilities, enhance communication with the public and improve the cooperation with the local communities.

After the 2014 BRICS Summit, China's investment in Latin America covered some super infrastructure projects, including canals, highways, railroads, hydroelectricity, mining and petroleum. The massive environmental impacts of these projects will put China into the frontline of international environmental protection. Therefore, Chinese investors should reconsider its domestic and international behaviors in terms of political rhetoric, media response, cultural adaptation and ecological responsibility.

CHAPTER 16

China's Role in Antarctic Marine Conservation

*Chen Jiliang**

Abstract

China officially joined the Commission for the Conservation of Antarctic Marine Living Resources (CCAMLR) in 2007 and has since been actively involved in the governance of Antarctic marine areas, as well as gradually carrying out fishing operations. This paper provides an overview of the current state of Antarctic marine conservation, an update on the negotiation for marine protected areas, and a summary on China's main concerns. It argues that China's position on Antarctic marine conservation should be built on the following two points: first, the overall development strategy for pelagic fishery should take into consideration the reality of global fishery degradation; second, polar policies should go hand-in-hand with China's political direction of ecological advancement, in which responsible decisions should be made based on the country's specific interests and restraints of reality.

Keywords

Commission for the Conservation of Antarctic Marine Living Resources (CCAMLR) – Southern Ocean – high sea marine protected area

1 Overview of Antarctic Marine Governance

The Antarctic krill fishing industry, which began in the 1960s and 70s, has raised serious concerns from signatory countries to the Antarctic Treaty and scientists. As a result, parties of the Antarctic Treaty initiated negotiations regarding the establishment of the Convention on the Conservation of Antarctic Marine Living Resources (CAMLR Convention). The mission of the CAMLR Convention

---

* Chen Jiliang is a researcher at Greenovation Hub. He has been participating in CCAMLR's negotiations as a representative from the NGO since 2012, and has been following the Commission's efforts on Marine Protected Areas.

is to ensure fisheries operate under the control measures of the Antarctic Treaty system, and will not cause serious negative effects on target species or ecosystems in the Southern Ocean. After the Convention was adopted in 1980, the signatories established the Commission for the Conservation of Antarctic Marine Living Resources (CCAMLR).

CCAMLR adopted ecosystem approach and precautionary approach in managing Antarctic marine conservation. These approaches are mainly reflected in the conservation principles specified in Article II, paragraph 3 of the Convention on the Conservation of Antarctic Marine Living Resources, including "maintenance of the ecological relationships between harvested, dependent and related populations of Antarctic marine living resources and the restoration of depleted populations to the levels defined in sub-paragraph (a) (its size should not be allowed to fall below a level close to that which ensures the greatest net annual increment)." In other words, CCAMLR is not only committed to the health of marine living resources being harvested, it is also committed to the health of living resources both upstream and downstream on the food chain, as well as the ecosystem as a whole. In addition, it manages fishing activities with cautious approaches.

2   Progress on Antarctic Marine Protected Area

2.1   *Current Status of Antarctic Marine Protection*

Since the Convention took effect, in order to achieve sustainable fishing and management, CCAMLR has implemented an efficient system of fishing restriction and inspection. Due to the decrease in fish populations, the Commission officially closed most of the finfish fisheries within the Convention Area, including those around the Antarctic Peninsula, South Orkney Islands, and in the Southern Indian Ocean. CCAMLR also implemented a series of measures to reduce illegal fishing for toothfish, including regularly patrolling many legal fisheries by inspectors from member countries. Apart from prohibiting gillnet fishing, CCAMLR also prohibited bottom trawling, as such practice harms the ocean floor. Meanwhile, the Commission implemented measures to protect vulnerable benthic habitats. CCAMLR's application of the precautionary approach is mainly by implementing catch limits. It also established temporary closed areas, such as Small-scale Management Units (SSMU) and Small-scale Research Units (SSRU).

In a sense, CCAMLR's Convention Area is the world's only marine area with strict management system established before large-scale fishing took place. The results are satisfactory after many years of efforts. For example, the incidental catch rate of seabirds was reduced to almost zero, and their ecosystems are relatively well protected. Compared to the gradual decrease in production year-over-year in most fishing areas around the globe, CCAMLR's management has become a best practice.

2.2   The Need to Create Marine Protected Area (MPA)

Nowadays the most widely used definition for Marine Protected Area derives from the International Union for Conservation of Nature (IUCN)'s definition for protected area: "a protected area is a clearly defined geographical space, recognized, dedicated and managed, through legal or other effective means, to achieve the long-term conservation of nature with associated ecosystem services and cultural values." In marine context, these areas are called Marine Protected Areas. This definition has been widely adopted by governments and NGOs. Therefore it is considered an international standard, on which individual countries' definitions of their own MPAs are based. The IUCN emphasizes that "spatial areas which may incidentally appear to deliver nature conservation but DO NOT HAVE STATED nature conservation objectives should NOT automatically be classified as MPAs."[1] Simply put, MPA is a designated area of ocean where human activities are restricted in order to protect the ecosystem of that area. Restricted human activities can include fishing, shipping, oil drilling, and tourism. CCAMLR is legally authorized to manage the fishing industry. Therefore it can only establish MPAs restricting fishing activity under this authorization. To protect the areas from other human activities, it needs to collaborate with relevant international organizations. For example, International Maritime Organization adopted the International Code for Ships Operating in Polar Waters in November 2014, establishing the safety and environmental standards for ships operating in polar waters.

Unfortunately, CCAMLR's excellent track record in conservation has often led people to question the need for creating an MPA in Antarctica. Based on the current proposal, it is necessary to establish an MPA in Antarctica for the following reasons.

First, there are many unknown factors regarding the influence of fishing activities. Despite the large population of species in the Antarctic marine ecosystem, the structure of food web is relatively simple.[2] Krill and toothfish

---

1   IUCN, "Guidelines for Applying the IUCN Protected Area Management Categories to Marine Protected Areas," IUCN, pg 9 (2012), http://cmsdata.iucn.org/downloads/uicn_categories amp_eng.pdf.
2   "Fantasizing about environmental warfare in the early 1960s, NATO scientists tried to imagine which links in ecosystems were vulnerable to manipulation.... It was a revelation to think that such a connection in the food chain was now targetable. But the reverse was also true,

are not only being heavily fished, but they are also at the center of the food web. We know very little about toothfish, except that their life span can go up to 50 years. Their foraging and spawning behaviors are largely unknown. Therefore it is necessary to protect the marine areas that are likely to become their spawning and foraging habitats. As for krill, despite efforts to disperse catch limits among a number of different areas, possible dense fishing activities nearshore still may have caused them to compete for food with penguins. There are concerns over the continuous increase in krill fishing in the future due to various reasons, the most important of which is the increasing demand for krill products, including the use of Omega 3 fatty acids in health supplements and aquaculture. The establishment of MPAs can redirect and alleviate some of the pressure brought on to the ecosystem by fishing, hence protecting some of the most important processes in the ecosystem such as foraging and spawning, all without lowering the catch limits.

Second, climate change has brought increasing uncertainties to the Southern Ocean. Due to the increase in offshore wind caused by global warming and ozone layer damage, winter sea ice west to the Antarctic Peninsula has been decreasing dramatically both in extent and age. Winter sea ice is precisely the "incubator" for krill. Sea ice decrease leads to krill population decrease in the summer of the following year, which affects various species that feed on krill. Meanwhile, fishing activities lasts longer and into late winter because of the sea ice shrinking, which in turn creates more pressure to the ecosystem. The increasing density of carbon dioxide also has led to ocean acidification, to the point that plankton shells corroded by acidic sea water have been found in the Antarctic. Even though an MPA cannot completely eliminate the negative influence of climate change, it can improve the self-recovery ability of the Antarctic ecosystem, providing a sanctuary for species to rebuild and revive.

Third, the Antarctic is the only large intact ecosystem in the world that remains wild, thanks to few human activities; hence it is an ideal nature laboratory. Protecting it can provide a reference point for studying the relative impacts of fishing and climate change on the overall status and health of ecosystems.

From the standpoint of the fishing industry, the Southern Ocean has indeed been well protected, compared to most of the other marine areas in the world. However, taking into consideration the special value of the Southern Ocean, the emphasis of the CAMLR Convention is not on sustainable fishing, but on

---

and ... the complexity of an ecosystem made any particular "link" less important, making the system less vulnerable." Jacob Darwin Hamblin, "Ecology Lessons from the Cold War," *New York Times*, June 1, 2013.

the conservation of ecosystems. From that perspective, setting up an MPA for the Southern Ocean will have tremendous benefits, which was why CCAMLR has decided to take that direction from the onset.

2.3 Negotiation Progress and China's Involvement

In response to the marine conservation goals set by the 2002 World Summit on Sustainable Development in Johannesburg, CCAMLR launched a series of discussions and workshops on Marine Protected Area in 2005. China officially became a member country of CCAMLR in 2007, and has supported the establishment of the South Orkney Islands Southern Shelf Marine Protected Area in 2009. However, China was strongly against the initial proposals to establish MPAs in Ross Sea and East Antarctica. In 2011, prior to drafting the official proposal for the two MPAS, China and Russia took a hard-line position during the discussion on "Conservation Measure 91–04: General framework for the establishment of CCAMLR Marine Protected Areas." After the two proposals were officially submitted in 2012, China and Russia refused to discuss them on the Commission's annual meeting that year, citing procedural reasons. As a result, the Commission held an additional special meeting on MPA in Bremerhaven, Germany in July 2013. During the special meeting, the scientific committee held in-depth discussions on the two proposals. Nonetheless, during the Committee's formal discussion, Russia questioned the legal rights of CCAMLR to establish MPAs. In the end the questioning led to no consequence. During the annual meeting in October 2013, Russia and Ukraine were once again strongly against the proposals going into resolution drafting stage, which had undergone many revisions (the proposed area of Ross Sea MPA had been reduced by as much as 40%). On the other hand, China's position had been relatively flexible. During the Antarctic Treaty Consultative Meeting (ATCM) in 2014, Russia openly raised its main concerns over the Antarctic MPA. The meeting then discussed these concerns, while China took on a constructive role. However, during the Commission's annual meeting in 2014, despite knowing Russia would strongly interfere with the passing of MPA resolutions, China openly and strongly questioned the necessity of establishing a protected area.

It is an innovative and experimental approach to establish a large-scale MPA on the high seas. Compared to MPAs under the governance of individual countries, the proposed Antarctic MPA has the following features: 1) it is comprised of vast areas with little data and few human activity; 2) it is managed by CCAMLR, but without detailed management plan or clear delegation of management costs and responsibilities; 3) the targeted areas in the official proposal correspond to sectors in the territorial claims originally made by the proposal initiating countries and could lead to skepticism from other countries.

With major reservations and no revision suggestion from China and Russia, the size of the proposed Ross Sea MPA had been reduced from 2.3 million km² in 2013 to 1.32 million km² in 2014, while the size of the proposed East Antarctica MPA had been reduced from 1.63 million km² in 2013 to 1.2 million km² in 2014. During the last 4 years, despite various efforts and the emergence of a few compromise solutions, the annual meetings and discussions have been largely inconclusive. Meanwhile, proposals to establish MPAs in Weddell Sea and west Antarctic Peninsula are in their respective planning phase. With regards to MPA negotiations, CCAMLR is faced with a difficult situation where new issues are piling on top of unresolved old issues.

3   China's Main Concerns over Antarctic Marine Conservation

China's position on Antarctic MPA is shaped by a number of factors, including geopolitical considerations, prospective interests in fisheries, and pragmatic concerns over the operation of MPAs.

3.1   *Geopolitical Considerations*

The proposed areas of the Antarctic MPA largely correspond to sectors in the territorial claims originally placed by the proposal initiating countries (i.e. New Zealand for Ross Sea proposal and Australia for East Antarctic proposal). This might be inevitable, as historical connection and geographical proximity have allowed the proposal initiating countries to be more active in certain parts of the waters; hence obtaining more data and having a bigger voice than other countries do. Nevertheless, this kind of correspondence worries non-initiating countries or countries opposed to the proposal. Even though initiating countries have been very careful to avoid causing suspicion of territorial claim—for example, putting the governance of protected areas entirely under the framework of CCAMLR—and the effort was enough to wipe away geopolitical skepticism from Latin American countries, as the result shows, it is not convincing enough for traditional rivals such as China and Russia.

3.2   *Interests in Fisheries*

As stated in "Opinions on Promoting Sustainable and Healthy Development of Pelagic Fishery," issued by the Ministry of Agriculture in 2012, "under the framework of CAMLR Convention, emphasize on developing Antarctic krill resources, conduct in-depth research, exploratory fishing, as well as processing and production on Antarctic marine resources ... and work towards the development and utilization of Antarctic krill resources in commercial scale."

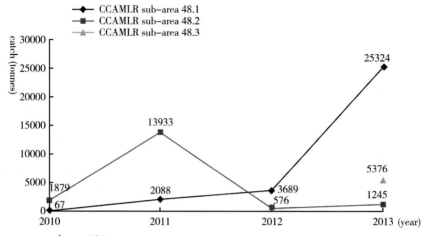

FIGURE 16.1   China's krill catch in the Antarctic
SOURCE: CCAMLR

China started exploratory fishing on Southern Ocean krill in 2009, and has essentially completed the transformation from exploratory fishing to commercial fishing. Chen Xuezhong, Director of East China Sea Fisheries Research Institute (ECSFRI) at the Chinese Academy of Fishery Sciences (CAFS) thinks that krill fishing in the Antarctic has important significance for the sustainable development of China's pelagic fishery. China has over 10 large fishing trawlers that were concentrated in the Northern Pacific region. As cod population declines, some of the trawlers moved to offshore Chile for mackerels. Whenever resource population fluctuate, the ships are left idle. In the future, Antarctic krill will hopefully become a backup resource for pelagic fisheries.[3] China is the world's largest fishing and aquaculture country. Therefore, it is also the largest importer of fishmeal. Krill harvesting might relieve the issue of low supply. Figure 16.1 shows China's Antarctic krill catch during exploratory fishing between 2010 and 2013. Each line of color represents the catch by sub-area.

---

3   Ren Quan, "China's First Pelagic Fishing Vessels Set Sail for Antarctic Krill Exploratory Fishing," December 4, 2009, http://news.sina.com.cn/c/2009-12-04/075519188530.shtml.

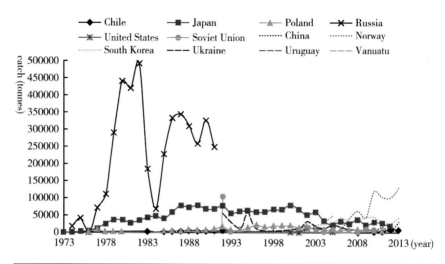

FIGURE 16.2   CCAMLR countries historical krill catch
SOURCE: CCAMLR

China is a latecomer in krill fishing. Figure 16.2 shows annual krill catch by country since the 1970s. As we can see, the highest harvest quantity is below 700,000 tonnes. China should aim to harvest one to two million tonnes, said Liu Shenli with the China National Agricultural Development Group Company.[4] This has drawn attention from those following the krill fishing industry.

The current MPA proposals do not change the existing catch limits. They only change the geographical range where fishing activities are allowed. In addition, the proposed MPA designation does not include area 48, where the majority of krill harvesting take place. The proposal for the Ross Sea MPA mainly concerns closing some of the toothfish fisheries. As shown in figure 16.3, the protected area in fact does not include areas with the highest catch. In other words, while planning the MPA proposal, proponent countries have already taken fishery interests into consideration to the greatest extent. However, for

---

4 Xie Yu, "Country steps up operations in Antarctic to benefit from krill bonanza," *China Daily USA*, March 4, 2015, http://usa.chinadaily.com.cn/epaper/2015-03/04/content_19716649.htm.

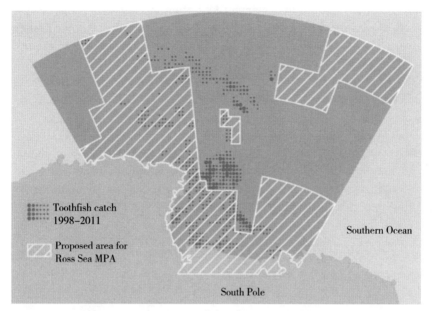

FIGURE 16.3   2012 Ross Sea MPA proposal
Note: dotted shades represent toothfish catch 1998–2011
SOURCE: ANTARCTIC OCEAN ALLIANCE

fishery latecomer countries, future fishing prospect is quite important as well. While developed countries comprehensively consider various interests and values including biodiversity, scientific research, landscape aesthetics, and the fishing industry, fishery latecomer countries and developing countries place more emphasis on the value of fishery which can be directly utilized.

3.3   *Research and Management Systems for Protected Areas*
China also has concerns over the operational practicality of research and management in the protected areas. This is because the management and research plans outlined in the current proposals are all based on the existing operational structure of CCAMLR, which reflects the idea of "maintaining existing system while improving in practice." In other words, unlike the management of Antarctic Specially Protected Areas and Specially Managed Areas, which is delegated to individual countries, CCAMLR manages MPAs as an entity. In the actual management practice, however, it is unclear how countries will share the additional law and regulation enforcement responsibilities and costs brought on by the MPA. Using South Orkney Islands Southern Shelf MPA as an example, there had been no research or management plan since its boundary was defined in 2009. It was only prior to the deliberation in 2014 did the

European Union drafted management and scientific research plans for discussion. Therefore, as indicated in a 2014 report on the South Orkney Islands Southern Shelf MPA,[5] from 2009 to 2014, the designation of the MPA did not lead to organized scientific researches focusing on the MPA. In the article "Analysis on the Problems with the 'General Framework for the Establishment of CCAMLR Marine Protected Areas,'" author Yang Lei argues that due to the lack of regulations on key components comprising the scientific basis for MPA designation, including baseline scientific data, scientific targets and standards system, and information collection and management system, the simple research and monitoring plan put forward by the proponent countries could hardly fulfill the responsibility of providing scientific basis and support for the monitoring, assessment, and review of MPAs.[6]

4 Building a Country of Marine and Polar Power through Ecological Advancement

China's position on Antarctic MPAs has always stood against the backdrop of the country's goal to become a marine and polar power, as well as to develop pelagic fishery. Nonetheless, ecological advancement is also a prominent goal of our times that cannot be ignored. The concept of ecological advancement was raised because learning from lessons in domestic economic development, the core leadership of the country came to the insight that the traditional economic development model based on the assumption of unlimited natural resources and aiming for infinite growth of material consumption has come to a dead end. However, it takes time for such understanding to effectively spread to all walks of life. The adoption is faster in areas such as energy conservation, emission reduction, and nature conservation, but could be slower in areas traditionally regarded as irrelevant to "environmental protection," such as marine and polar development.

It is China's established principle to build a country of marine power and "contribute to the peaceful uses of Antarctic resources by mankind." When referencing the principle for guidance on current issues, it is necessary to put it against the backdrop of promoting sound ecology.

First of all, there must be a rational and long-term overall strategy for pelagic fishery addressing general trends in the global fishing industry. As

---

5 Document ID: CCAMLR, sc-xxxiii-bg-19.
6 Yang Lei, "Analysis on the Problems with the 'General Framework for the Establishment of CCAMLR Marine Protected Areas,'" *Chinese Journal of Polar Research*, 26, 532.

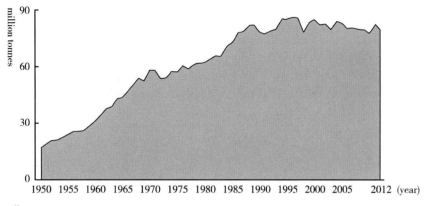

FIGURE 16.4   The trend in global marine fisheries
SOURCE: THE STATE OF WORLD FISHERIES AND AQUACULTURE 2014 BY FOOD AND AGRICULTURE ORGANIZATION, PAGE 5, HTTP://WWW.FAO .ORG/3/A-I3720E.PDF

reported by the Food and Agriculture Organization of the United Nations, after almost 40 years (1955–1990) of rapid increase, the total catch of marine fishing industry is basically at its peak level and currently going steady while on a gradual downward slope (figure 16.4). China began expanding its pelagic fishery at the end of the 40-year increase. This means in terms of quantity, the rapid development of China's pelagic fishery did not tap into new resources. It only transferred some of the fishing capacity from other countries to China. If this trend continues, China will be taking a lead role in global pelagic fishing and be carrying relevant responsibilities, for which China must be prepared. By being prepared, we mean having a plan for the next steps once the fishing capacity stops increasing. One might say we could go to other countries' exclusive economic zones or the high seas when domestic capacity max out, but what if capacity at all these other places maxed out as well? Therefore, we need to backward induct the current set of policy from a strategic level, enabling bigger and better growth. Better growth does not refer to the increase in catch per unit effort (CPUE). It refers to stronger fishery management capability. This requires advanced concepts, such as the ecosystem principle and the precautionary principle. It also requires strong capabilities in policy implementation and law enforcement, such as combating illegal fishing. The end goal would be to separate production value from catch quantity, creating a win-win situation for both ecosystem and economy. Fishery countries hope to expand Antarctic krill harvesting, while environmental organizations unanimously oppose it for one same reason: existing fishing activities in other marine areas have caused problems. Fishing industry argues that the problems reflect nature's limits, and can only be resolved by expanding fisheries to other places, while environmental organizations argue that the problems came from the management of fishing industry itself. Unless there is fundamental change to the development model focusing on the increase in productivity, raising krill catch limits in the Southern Ocean would only lead to the repetition of mistakes made on other marine areas.

Secondly, the principle "contribute to the peaceful uses of Antarctic resources by mankind" should take on new meaning based on the characteristics of our times. In particular, the word "use" has special meaning in the context of ecological advancement and under the Antarctic Treaty system. The 30 years of Antarctic exploration coincides with 30 years of reform and opening up, globalization, as well as sustainable development becoming a global agenda. During these years, the understanding of "use" has undergone tremendous changes as well. Although the concept "limits of growth" came forward as early as 1972, it only became globally accepted in the last 30 years. In particular, after the World Summit on Sustainable Development in Johannesburg, the concept is integrated with "sustainable development." Simply put, limits of growth mean human's uses of Earth's material resources need to be limited to a certain range in order to ensure the speed of material consumption is equal or less than the speed of self-recovery. Both "rational use" and "sustainable use" express the same idea, which is clearly represented by Article III of the Convention on the Conservation of Antarctic Marine Living Resources. The idea is based on the understanding that the economic activities of mankind are a sub-system of ecosystem, not the other way around. Because this understanding directly jeopardizes the political agenda of "economic development comes first," it encountered many obstructions and barely entered the view of mainstream economics after a long struggle. While limits of growth is a limitation on the traditional concept of "use," discoveries in the science world has extended this traditional concept. Development in ecology and lessons from environmental damage in the last few decades have led us to realize resources that can be privatized and monetized are not the only things that can be utilized. Human beings have always been enjoying free services from the ecosystem, such as clean air, pure water, and biodiversity. Once these services are lost, we have to pay the price with money. Therefore, providing ecosystem service and conducting scientific research both belong to "use" in its broad sense. In short, protection means to restrict "use" in its narrow sense while the goal of protection is to achieve "use" in its broad sense.

5 Conclusion

In summary, China's position on Antarctic marine conservation agenda should be built on the following two points: first, the overall development strategy for pelagic fishery should take into consideration the reality of global fishery degradation; second, polar policies should go hand-in-hand with the idea of ecological advancement. Subject to the country's specific interests and restraints from reality, China will make responsible decisions on issues regarding such global public goods as Antarctic marine living resources.

Last but not least, as an environmental organization, we think each country involved in the governance of the Antarctic need to ask itself a question: how would I treat the Antarctic if it was my own territory? To answer this question, one must first answer the question "how are my own territories being treated right now?" The answers to these two questions determine if the country can claim the moral high ground in Antarctic governance. Negotiations can resolve most issues on interest. While the designation of the Antarctic MPA needs active participation from China, we hope as with climate change negotiations, compromises can be reached without reluctance but with eagerness.

CHAPTER 17

Real-time Information Disclosure under Blue Sky Roadmap II

Jointly released by the Institute of Public & Environmental Affairs; Society of Entrepreneurs & Ecology (SEE); the Institute of Environmental Policy and Planning, Renmin University of China; Friends of Nature; EnviroFriends; Nature University

In 2013, China's urban air pollution drew wider attention in society. Many areas were shrouded in haze; in particular, the densely populated regions, including Beijing, Tianjin, Hebei, some areas in Shandong and Henan, were often covered by haze for days, while severe haze in Northeastern China and the Yangtze River Delta Region seriously affected the local areas. A blue sky is still far away from us.

How can we recover a blue sky? The Institute of Public & Environmental Affairs released a report in December 2011, calling for moving toward the goal of restoring a blue sky step-by-step in the sequence of monitoring and releasing, early warning and emergency response and identification of pollution sources, and it focused on emission reduction. In comparing it to a roadmap, we can find that in 2013, China made important progress in information release, early warning and emergency responses, while there were opportunities for making historical breakthroughs in the key link–identification of pollution sources.

As shown by the AQTI index evaluation of the comprehensiveness, timeliness, completeness and user-friendliness of information disclosure, the level of disclosure of the information regarding the quality of the air of nearly 100 cities continued to increase significantly, with the average score rising from 21.5 in 2012 to 58.8 in 2013. As of January 2, 2014, 179 cities disclosed the information regarding the quality of their air on a real-time basis, residents can get access to real-time data on air quality through a computer, even on a mobile phone.

As the real-time disclosure gave prominence to the degree of pollution severity, many local authorities developed contingency plans for heavy pollution weather. Among these emergency measures, urging schools to stop outdoor activities and forcing enterprises to reduce emissions were the key measures for protecting public health and preventing worsening of pollution. According to the report, with long-term coordination efforts, these key measures started to be implemented under heavy pollution weather; however, there is no comprehensive disclosure, so it is often hard to confirm the status and effect of implementation.

Monitoring, releasing, early warning and emergency responses are important; however, in order to recover a blue sky, it is necessary to reach large-scale emission reduction, while pollution sources must be identified before emission reduction can be achieved. Based on a relevant analysis, we believe that the current level of pollution has shown obvious regional characteristics, so the identification of pollution sources must be extended to the regional level. According to relevant research, in such regions as Beijing, Tianjin, Hebei and the Yangtze River Delta Region, high energy-consuming industries are concentrated, and coal consumption is huge, so a number of point sources with huge emissions should be controlled first.

In our opinion, in order to identify and supervise pollution sources within the regions, it is essential to extend information disclosure from PM2.5 to pollution sources. As from Yabuli Forum in February 2013, environmental protection organizations and entrepreneur organizations worked together to promote the disclosure of pollution source information, which received active responses from Beijing, Hebei and other regions. The Ministry of Environmental Protection issued relevant regulations on July 30, 2013, requiring various provinces to build platforms for real-time information disclosure of online monitoring data.

Such provinces as Shandong, Zhejiang and Hebei have stayed ahead in real-time information disclosure of online monitoring data. We hold that their good practice is conducive to satisfying the public's right to know and identifying the main pollution sources within the regions. At the same time, regrettably, the online platforms of important provinces and municipalities, including Tianjin, Guangdong and Hunan, did not disclose information.

Based on a preliminary analysis of these online data, we can find that a number of thermal power, iron and steel enterprises in Northern China, including Shandong and Hebei, severely exceeded the emission standard, even some major emitters continuously committed emissions beyond the standard in some periods with heavy pollution at the local level.

Real-time information disclosure contributes to identifying pollution sources within the regions. A comparison of online data reveals a significant gap in the emission scale of industrial pollution sources among regions. Taking some major enterprises in Shandong, Hebei and Beijing during October—December 2013 as an example, the total oxynitride emission from 8 enterprises in Shandong and Hebei was 37 and 30 times greater than that from 8 major enterprises in Beijing; these pollution sources should be the focus of emission reduction efforts.

It should be pointed out that in advance, Shandong has imposed stricter emission standards on major industries, including thermal power, iron and steel industries; also Hebei has carried out stricter emission standards for the iron and steel industry. However, the emission standards in major provinces and municipalities, such as Jiangsu, Zhejiang and Liaoning, including Hebei's emission standards for the thermal power and cement industries and Tianjin's oxynitride emission standards, need to be raised.

The extensive haze in 2013 has also aroused the most aggressive government action plan. On September 10, 2013, the State Council released the *Action Plan for Air Pollution Prevention and Control*, calling for taking five years to significantly improve the quality of the air and another five years to remove heavy pollution weather by adopting ten measures concerning pollution control and adjustment of the industrial and energy structures.

In our opinion, given a staggering number of pollution sources, relevant emission reduction plans must focus on key points and start from controlling industrial and coal-fired enterprises. According to calculations, if some enterprises in key areas of Shandong and Hebei can meet emission standards, their emissions of the most critical pollutants, including oxynitride, can be greatly reduced.

The *Action Plan for Air Pollution Prevention and Control* and the emission reduction measures adopted by local authorities have inevitably moved the enormous vested interests, and the challenges for implementation cannot be underestimated. The people from various sectors of society who pay attention to air pollution control cannot merely wait for a blue sky. We call on the government, courts, enterprises, media, environmental protection organizations and citizens to play their respective roles and seize the historical opportunities presented by real-time disclosure of pollution source information, proceed from supervising enterprises in meeting emission standards to jointly promoting pollution emission reduction and as soon as possible to dispelling the haze in cities.

In 2013, China's urban air pollution drew wider social attention. On the one hand, nearly 100 cities have started real-time disclosure of the monitoring information concerning such pollutants as PM2.5; the central and local governments have unveiled huge action plans for emission reduction; on the other hand, air pollution is still getting worse and a large number of cities have been frequently hit by haze; in particular, the densely populated Beijing, Tianjin, Hebei, and some areas of Shandong and Henan were, for several times, subject to haze for days, while severe haze exerted a serious impact on Heilongjiang, Jilin and the Yangtze River Delta Region.

Citizens' anxiety has continually grown amidst haze. A blue sky is still out of reach for us. How can we recover our blue sky?

For China, amidst rapid industrialization and urbanization, it is hard to control air pollution overnight. The Institute of Public & Environmental Affairs released a report in December 2011, calling for moving toward the goal of restoring a blue sky step by step in the sequence of monitoring and releasing, early warning and emergency response, identification of pollution sources, and it focused on emission reduction.

Specifically, the blue sky roadmap covers the following four steps:

> Step One: Extend the disclosure of air quality information, comprehensively and promptly release monitoring data to society;
> Step Two: Give the corresponding health early warning to the general public, take vigorous emergency measures to mitigate heavy pollution;
> Step Three: Identify the main sources of pollutant emissions, determine the focus of emission reduction;
> Step Four: Develop the well-targeted plan and timetable to substantially reduce emissions;

This report presents the current progress against the roadmap, aims at finding out the key bottlenecks which urgently need to be tackled, so as to seize the key points for making large-scale emission reduction as soon as possible.

According to the information disclosure made by 113 key cities for environmental protection (Table 17.1), the Institute of Public & Environmental Affairs conducted the third Air Quality Information Transparency Index (AQTI) evaluation to confirm whether the disclosure of air quality information by various cities was comprehensive, prompt, complete and user-friendly.

TABLE 17.1 The results of the AQTI evaluation of 113 key cities for environmental protection

Ranking	City	Score in 2013	Score in 2012	Ranking	City	Score in 2013	Score in 2012
1	Beijing	77.4	64.8	11	Wenzhou	73.2	24.0
2	Dongguan	76.8	69.0	11	Shaoxing	73.2	37.8
3	Nanjing	76.4	56.0	11	Fuzhou	73.2	15.0
3	Suzhou	76.4	55.2	11	Yantai	73.2	18.6
3	Chongqing	76.4	30.6	11	Wuhan	73.2	47.4
6	Ningbo	76.0	9.0	11	Chengdu	73.2	42.6
7	Dalian	74.8	54.8	11	Kunming	73.2	16.8
7	Qingdao	74.8	28.6	20	Xiamen	71.6	43.0

TABLE 17.1   The results of the AQTI evaluation of 113 key cities (*cont.*)

Ranking	City	Score in 2013	Score in 2012	Ranking	City	Score in 2013	Score in 2012
7	Guangzhou	74.8	16.2	20	Ji'nan	71.6	13.8
10	Jiaxing	74.2	76.0	22	Shanghai	71.0	50.2
11	Tianjin	73.2	33.6	23	Taizhou	70.2	22.0
11	Hangzhou	73.2	20.4	24	Changzhou	70.0	39.6
24	Nantong	70.0	44.2	60	Changzhi	61.2	15.8
24	Lianyungang	70.0	28.8	60	Shantou	61.2	16.2
24	Yichang	70.0	18.6	62	Hohhot	61.0	8.4
28	Huzhou	68.4	18.0	62	Shenyang	61.0	11.4
28	Zibo	68.4	11.4	62	Changchun	61.0	11.4
28	Zaozhuang	68.4	13.8	62	Yancheng	61.0	18.6
28	Weifang	68.4	0.0	62	Nanchang	61.0	18.6
28	Jining	68.4	0.0	62	Changsha	61.0	13.8
28	Tai'an	68.4	19.2	62	Xiangtan	61.0	13.8
28	Weihai	68.4	7.2	62	Urumqi	61.0	14.4
28	sunshine	68.4	0.0	70	Datong	57.6	17.4
36	Xi'an	68.2	38.6	70	Yangquan	57.6	17.6
37	Foshan	67.6	64.8	70	Linfen	57.6	18.2
38	Shenzhen	67.4	75.0	73	Beihai	53.8	16.2
39	Zhuhai	66.0	56.4	74	Baotou	52.2	8.4
39	Zhongshan	66.0	64.6	74	Ma'anshan	52.2	11.4
41	Taiyuan	65.2	25.4	74	Quanzhou	52.2	15.6
41	Zhengzhou	65.2	13.8	74	Shaoguan	52.2	14.4
41	Kaifeng	65.2	14.4	74	Liuzhou	52.2	14.4
44	Shijiazhuang	64.6	19.2	74	Guilin	52.2	13.8
44	Tangshan	64.6	19.8	74	Mianyang	52.2	9.0
44	Qinghuangdao	64.6	5.4	74	Yibin	52.2	15.6
44	Handan	64.6	11.4	74	Baoji	52.2	16.8
44	Baoding	64.6	18.6	74	Xianyang	52.2	13.8
44	Harbin	64.6	27.0	84	Erdos	50.4	4.2
44	Wuxi	64.6	31.8	84	Anshan	50.4	11.4
44	Xuzhou	64.6	22.2	84	Tongchuan	50.4	19.8
44	Yangzhou	64.6	31.2	84	Yan'an	50.4	9.6
44	Hefei	64.6	30.6	88	Wuhu	46.6	14.4
44	Zhuzhou	64.6	16.8	88	Jingzhou	46.6	19.2

TABLE 17.1 The results of the AQTI evaluation of 113 key cities (*cont.*)

Ranking	City	Score in 2013	Score in 2012	Ranking	City	Score in 2013	Score in 2012
44	Nanning	64.6	38.6	90	Jinzhou	38.0	14.4
44	Guiyang	64.6	19.8	91	Chifeng	37.6	11.4
44	Lanzhou	64.6	14.4	91	Fushun	37.6	22.8
44	Xining	64.6	14.4	91	Benxi	37.6	0.0
59	Yinchuan	62.4	15.0	91	Jilin	37.6	11.4
91	Qiqihar	37.6	11.4	91	Zhangjiajie	37.6	11.4
91	Daqing	37.6	11.4	91	Zhanjiang	37.6	14.4
91	Mudanjiang	37.6	9.0	91	Panzhihua	37.6	11.4
91	Jiujiang	37.6	16.2	91	Luzhou	37.6	11.4
91	Luoyang	37.6	14.4	91	Zunyi	37.6	11.4
91	Pingdingshan	37.6	14.4	91	Qujing	37.6	0.0
91	Anyang	37.6	18.0	91	Jinchang	37.6	0.0
91	Jiaozuo	37.6	4.2	91	Shizuishan	37.6	15.6
91	Yueyang	37.6	13.8	91	Karamay	37.6	11.4
91	Changde	37.6	14.4				

CHAPTER 18

An NGO Review of China's Carbon Market

Greenovation Hub

The trading of carbon emissions permits is one of the policy tools for adopting a market means to achieve energy conservation and emission reduction at lower costs; it is an important embodiment of policy innovations and emission reduction improvement in China during the period of the 12th Five-Year Plan. In the second half of 2011, the National Development and Reform Commission started to pilot carbon trading in seven provinces and municipalities. The year 2013 was the year for kicking off pilot carbon trading, producing a major demonstration effect on the future first-ever implementation of China's carbon price policy. Given the developmental stage and level of the Chinese carbon market, based on fully drawing upon international experience and the opinions from experts, Greenovation: Hub has worked on building a framework system for evaluating the Chinese carbon market. The evaluation system covers four main fields, including mechanism design, mechanism execution, market performance, information transparency and participation of stakeholders. For the four main fields, specific indicators have been developed to analyze and evaluate the operations of carbon trading.

This report summarizes the publicly available information concerning seven pilot areas as of November 5, 2013. As the public data obtained are not sufficient, this report provides a qualitative analysis of pilot priorities. The analysis uncovers the potential risks and challenges in the Chinese carbon market: the quality of emission data is inherently deficient; the setting of total quantity is too loose; the floatability of the total quantity, the connection between the reserved growth spaces and the energy and pollution control measures of other countries needs to be improved. The quota allocation methods adopted on a pilot basis are dominated by the grandfathering method based on free allocation and historical emissions, while in particular areas, beneficial attempts are being made by adopting the benchmarking and auctioning methods. The electric power industry is facing structural challenges, and the inclusion of the upstream and downstream sections in the operations of the carbon market still needs to be reexamined. In the existing practice, the legal basis and intensity of punishment are not sufficient; the monitoring reporting verification (MRV) mechanism, third-party qualification management and market risk control should be further improved; explicit regulations should be made for the use of economic income from the carbon market.

According to international experience and the current pilot work in China, the development of the carbon market is a process in which errors are constantly discovered and corrected; it is also a process characterized by continuous participation of stakeholders, experimentation of methods, summarization of experience and carrying out improvements. As policies are imperfect, it is more necessary to make trading information public and transparent. Prompt and accurate disclosure of information is exactly the important precondition for promoting the market to discover and correct problems. However, so far, the Chinese carbon market is unsatisfactory in information disclosure and data accuracy, there are no clear social supervision mechanisms and routes. As an important supplement to total quantity control and to the trading system, Chinese Certified Emission Reduction (CCER) brings the opportunities for the redevelopment of China's carbon offset projects and also poses challenges. In order to achieve a balance in demand and supply of carbon offset projects, it is essential to develop a rational control plan to ensure that the carbon offset mechanism can take into account various factors, including sustainable development and technology transfer, while promoting carbon emission reduction in an effective, efficient and low-cost way.

Regional carbon trading was put into practice in seven major pilot areas as from 2013; the top-level design of the national system of carbon trading has also been carried out; this will gain valuable experience for establishing a unified national carbon market in China in the future. Moreover, the carbon tax policy is awaiting political decisions. The policy development and building of a mechanism in the coming years are the key to the formation of China's carbon price system. Carbon pricing and the formation of an effective market aim at reducing greenhouse gas emissions and are also crucial for a global response to climate change. This process entails comprehensive thinking and practice, diversified participation and discussion, effective market supervision and social supervision for ensuring a fair market and an effective emission reduction. However, the current policy discussions are conducted in a small circle and a closed state; different stakeholders involved in the market mostly reflect the appeals and concerns involving business interests. As a non-governmental organization dedicated to environmental and climate protection, Greenovation Hub is carrying out the first evaluation of the carbon market in China. Greenovation Hub will combine systematic data collection with objective analysis to help conduct an inclusive and open discussion regarding the carbon market policy with the participation of various parties and based on the optimal experience both domestically and abroad, and will provide a reliable basis for ensuing independent non-governmental analysis and supervision to avoid mistakes and guarantee smooth policy-making and execution.

With the people's increasing attention to energy security and environmental issues and the mounting pressure on global response to climate change, China has intensified its efforts in energy conservation and emission reduction. The 12th Five-Year Plan specifies the goals of reducing carbon intensity and energy intensity by 17% and 16% in 2015 over 2010. Although the government-led administrative restriction on emissions produced a certain effect and reduced energy intensity by 19.1% during the period of the 11th Five-Year Plan, its limitations and problems are obvious. By contrast, the market-based policy is more able to push forward emission reduction action in a cost-effective way and seek a balance between economic development and response to climate change. Therefore, the establishment of a carbon market is an important attempt made by the Chinese Government in energy conservation and emission reduction.

In the second half of 2011, the National Development and Reform Commission initiated pilot carbon trading in seven provinces and municipalities, including Beijing, Tianjin, Shanghai, Guangdong, Shenzhen, Chongqing and Hubei. As carbon trading is a new mode which is about to be put into practice, every pilot effort faces big challenges, it varies with different regions in mechanism design, execution and market performance. Besides these differences, mechanism design is also subject to common challenges, including those involving data authenticity and accuracy, the setting of total quantity, allocation scheme, trading mode and the capability of participants. From 2012 to early 2013, Beijing, Shanghai, Guangdong, Tianjin and Hubei finalized or released their implementation plans for pilot carbon work.

The carbon market information collected and used in this chapter is the information obtained through public channels as of November 5, 2013. Table 18.1 summarizes the basic information on the pilot efforts in various areas. The results of our calculations based on the existing information on these pilot areas are similar to those from the World Bank and The Climate Group; the total carbon dioxide emission covered by these seven pilot areas was 650–700 million tons, equivalent to 9%–10% of China's current total amount of carbon dioxide emissions.

At the same time, related work was initiated under the national carbon trading system and relevant parts—including register and the mechanism for verifying the monitoring report—with the help and cooperation of the World Bank, EU governments and the Australian Government. The latest developments and the future three-year research plan can be found in the World Bank's Partnership for Market Readiness Project—China Carbon Market Initiative—approved in March 2013. In October 2013, the General Office of the National Development and Reform Commission issued the first *Accounting Method and Reporting Guidelines for Greenhouse Gas Emissions from 10 Industrial Enterprises (Trial)* which serve as the reference for carrying out trading of carbon emissions permits, building the enterprise greenhouse gas emission reporting system and improving the statistical accounting system for greenhouse gas emissions.

Moreover, the carbon offset projects constitute an important part of the Chinese carbon market. The National Development and Reform Commission issued the *Interim Measures for the Management of Greenhouse Gas Voluntary Emission Reduction Trading* in June 2012, specifying the rules for filing, development and management of China's voluntary carbon offset projects. In early March 2013, 52 methods specified in the first filing list of China's greenhouse gas voluntary emission reduction methodologies were released. In October 2013, the Chinese Certified Emission Reduction (CCER) trading information platform was launched, with the first 11 projects made public at the examination stage (October 24, 2013–November 8, 2013). On November 4, 2013, the second filing list of China's greenhouse gas voluntary emission reduction methodologies was released, including carbon sequestration afforestation and bamboo afforestation carbon sequestration.

The year 2013 was crucial for building the Chinese carbon market; in 2013, quota allocation, the design of the relevant mechanism and the trading platform construction were successively completed with regard to local pilot carbon trading, and trading was carried out (Table 18.1); pilot carbon trading brought about the first result. The top-level design of the national carbon trading system has been started, while the carbon tax policy is awaiting a political decision; the development of the carbon market in the coming years and the construction of its mechanism are crucial for forming China's carbon pricing system. However, so far, policy discussions and design are still conducted in a small circle and a closed state; different participating stakeholders have different appeals. An inclusive and open policy discussion, with participation of various parties and based on optimal domestic and foreign experience, was conducted, and on this basis, an independent social supervision mechanism gradually took shape, which is important for and beneficial to avoiding mistakes and guaranteeing smooth policy-making and execution. A multi-perspective and multi-dimensional evaluation system offers a solid foundation for arousing and promoting active discussion; this was the original purpose for preparing this report.

TABLE 18.1 Basic information on the implementation plan for local pilot carbon trading (as of November 5, 2013)

	Beijing	Shanghai	Guangdong
Release of implementation plan	April 10, 2012	August 16, 2012	September 7, 2012
Mechanism type	Mandatory	Mandatory	Mandatory
Type of total quantity	Absolute	Absolute	Absolute
The scope of mechanism (industry)	Electric power generation, thermal, manufacturing industry, large-scale public buildings, etc. (about 600)	High energy-consuming industries, aviation, seaport, railway, commercial buildings, etc. (197)	9 high energy-consuming industries, including electric power (827)
Energy intensity and emission reduction goals during the period of the 12th Five-Year Plan (% of quantitative value in 2010)	-17	-18	-18
Carbon intensity and emission reduction goals during the period of the 12th Five-Year Plan (% of quantitative value in 2010)	-18	-19	-19.5
Access threshold (1,000t CO_2/year)	10	20	20
Benchmark year (year)	2009–2011	2010–2011	2010–2012
Covered CO_2 emissions (10,000t)	4200	11000	21420

Tianjin	Hubei	Shenzhen	Chongqing
February 5, 2013	February 18, 2013	October 30, 2012	Approved by the National Development and Reform Commission in August 2013, not made public to society
Mandatory	Mandatory	Mandatory	Mandatory
Absolute	Absolute	Relative	Absolute
Iron and steel, chemical, electric power supply, heat supply, petrochemical, oil and natural gas exploitation and building	Iron and steel, chemical, cement, automobile manufacturing, electric power, non-ferrous, glass, papermaking etc. (about 153)	635 industrial enterprises and more than 200 large-scale public buildings	Electrolytic aluminum, ferrous alloy, calcium carbide, caustic soda, cement, iron and steel etc. (about 300)
-18	-16	-19.5	-16
-19	-17	-21	-17
0	~16	5	20
present-2009	2010–2011	2009–2011	2008–2011
800	12400	3173	5600

TABLE 18.1 Basic information on the implementation plan for local pilot (*cont.*)

	Beijing	Shanghai	Guangdong
Proportion in regional total emissions (%)	42	45–55	42
Quota allocation	Free initial quota during the period 2013–2015. A small part is taken as the reserved quota and may be dealt with through auctioning	Free initial quota during the period 2013–2015. Grandfathering method (with such correction factors as efficiency level taken into account). Auctioning will be considered in the future	Free initial quota during the period 2013–2015. Grandfathering method is dominant, supplemented by auctioning
	One-off allocation	One-off allocation	Annual allocation
MRV guidelines	MRV guidelines are in the making, supervised by the Beijing Development and Reform Commission and certified by the third party	MRV guidelines were finalized in the first half of 2013, with specific guidelines available for industries	MRV guidelines are in the making

Tianjin	Hubei	Shenzhen	Chongqing
50	35	38	35–40
Free initial quota during the period 2013–2015. Grandfathering method is dominant, supplemented by benchmarking. Total emission control in the next year is adjusted according to the certified emissions in the previous year	Free initial quota during the period 2013–2015. 5% is taken as the reserved quota and used for market regulation, while 15% is used for new enterprises and facilities	Free initial quota during the period 2013–2015. Quotas are pre-allocated and subsequent adjustment is made to determine the result	Unknown
Annual allocation	Annual allocation	Annual allocation	One-off allocation
MRV guidelines are in the making, supervised by the Beijing Development and Reform Commission and certified by the third party	MRV guidelines are in the making. The enterprises annually consuming 8,000t standard coal are mandatorily required to submit a carbon emissions report, gradually included into carbon trading on a batch-by-batch basis	The *Guidelines of Shenzhen City for Quantification and Reporting of Greenhouse Gases from Organizations* and the *Guidelines of Shenzhen City for Verification of Greenhouse Gas Emission by Organizations* were released in January 2012; the *Standards and Guidelines of Shenzhen City for Quantification and Reporting of Greenhouse Gas Emissions from Buildings* were released in April, 2013	MRV guidelines are in the making

TABLE 18.1 Basic information on the implementation plan for local pilot (*cont.*)

	Beijing	Shanghai	Guangdong
Carbon offset mechanism	CCER is permitted to be used; carbon offset under the Panda Standard may be permitted to be used. The cap for using carbon offset is 5% of annual carbon emissions	CCER is permitted to be used; the cap for using carbon offset is 5% of annual carbon emissions	CCER and voluntary emission reduction certified by Guangdong Province can be included in the carbon emission trading system. The cap permitted to be used is 10% of annual carbon emissions
Carbon storage (Yes/No)	Yes (before 2015)	Yes (before 2015)	Yes (before 2015)
Carbon lending and borrowing (Yes/No)	No	No	No
Trading platform	China Beijing Environment Exchange	Shanghai Environment and Energy Exchange	China Emissions Exchange
Performance and punishment	Not disclosed, unknown	Where a unit included in the quota management does not fulfill the reporting obligation, it will be subject to a fine above 10,000 yuan and below 30,000 yuan; where it does not accept verification action according to regulations, it will be subject to a fine above 10,000 yuan and below 50,000	Where carbon emissions exceed emission quota, a fine being three times the average market price of carbon emissions which violates regulations will be imposed

Tianjin	Hubei	Shenzhen	Chongqing
CCER is permitted to be used; the cap is 10% of annual carbon emissions	CCER within Hubei Province is permitted to be used; the cap is 10% of annual carbon emissions	CCER is permitted to be used; the cap is 10% of annual carbon emissions	CCER is permitted to be used; the cap is 8% of annual carbon emissions
Yes (before 2015)	Yes (before 2015 and subject to restrictive conditions)	Yes (before 2015)	Yes (before 2015)
No	No	No	No
Tianjin Emissions Exchange	Wuhan Optics Valley United Property Rights Exchange	China Emissions Exchange	Chongqing Assets and Equity Exchange
Not disclosed, unknown	Where there is an unpaid difference, a fine being three times the average market price of the current year's carbon emission quota will be imposed, deduction is made from the next year's allocated quota on a doubled basis	Where carbon emission exceeds emission quota, a fine being three times the average market price of carbon emissions which violates regulations will be imposed. Where an enterprise whose emission is subject to control does not submit a report on schedule or does not submit a sufficient quota on schedule, the verifying body does not conduct objective	Not disclosed, unknown

TABLE 18.1 Basic information on the implementation plan for local pilot (*cont.*)

	Beijing	Shanghai	Guangdong
Performance and punishment (*cont.*)		yuan, as appropriate; where it does not fulfill the quota settlement and payment obligation, it will be subject to a fine above 50,000 yuan and below 100,000 yuan, as appropriate. Where the third-party verifying body violates regulations and the circumstance is serious, it is not permitted to engage in this city's carbon emission verification activity for a period of three years, and will be subject to a fine above 30,000 yuan and below 50,000 yuan	

Tianjin	Hubei	Shenzhen	Chongqing
		verification or divulges enterprise information, the exchange does not perform its duties, a fine between 10,000 yuan and 500,000 yuan will be imposed, and the criminal act will be subject to criminal liability	

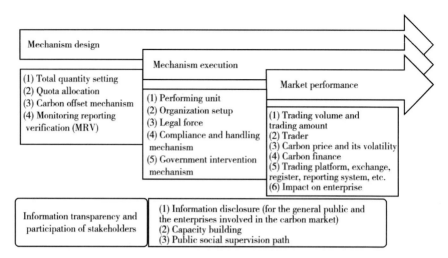

FIGURE 18.1   The framework of the Chinese carbon market evaluation system

This report starts with introducing the research framework and key indicators of the Chinese carbon market evaluation system and then adopts the system framework to evaluate two chosen typical pilot cities, Shenzhen and Shanghai, and on this basis raises questions and offers suggestions. Given the relative independence and importance of the carbon offset projects and their internal connection with China's carbon trading practice and development, especially the clean development mechanism and process, the carbon offset projects are analyzed and evaluated. Finally, this report will take the non-governmental perspective of global climate protection to offer suggestions for the issues concerning future market development, e.g. the efficiency of the Chinese carbon market, the effectiveness of emission reduction, the fairness for traders and procedural transparency.

TABLE 18.2 Overview of the non-governmental evaluation system for the Chinese carbon market in 2013

Main evaluation item	Evaluation subitem	Indicator	Environmental impact	Social impact	Economic impact
A. System design	1. Total quantity setting	The degree of strictness, including: what is the total quantity (the initial allocated quantity for existing enterprises)? How to treat increment (is the growth space reserved for economic development)?	▓		▓
		Flexibility: can adjustment be made (Australian mode)? What is the impact on the degree of the strictness of the goal? Is it absolute total quantity or relative total quantity?	▓		
		Are there goals for sub-industries?	▓		
		How can the enterprises which are new entrants be treated? Will it lead to unfairness?		▓	▓
		The scope of inclusion: generally, the industries with heavy emissions and low management costs are included, while it is relatively difficult to include the transportation and construction industries; many local authorities include these industries on a pilot basis, how can we operate?			▓

TABLE 18.2 Overview of the non-governmental evaluation system (*cont.*)

Main evaluation item	Evaluation subitem	Indicator	Environmental impact	Social impact	Economic impact
		What is the proportion of the total quantity in regional total emissions?			
	2. Quota allocation	The accuracy of historical data examines: is verification conducted? What about completeness and quality?			
		Allocation method and allocation mode: grandfathering or benchmarking; what is the proportion of freely allocated quantity? What is the force for driving enterprises to reduce emissions? Do methods take into account the efficiency level (early emission reduction) and industrial benchmarking or merely take historical emissions as a reference?			
		At the enterprise level: compare enterprise goals with the energy conservation goals of 10,000 enterprises			
		Does quota allocation involve the process of bargaining? Do state-owned enterprises enjoy more advantages?			

TABLE 18.2 Overview of the non-governmental evaluation system (*cont.*)

Main evaluation item	Evaluation subitem	Indicator	Environmental impact	Social impact	Economic impact
	3. Carbon offset mechanism	Supply and demand of carbon offset: the quantities of project and indicator supply, the relation between market supply and demand; carbon offset prices (ceiling price, floor price); proportion, source, type, etc.			
		Carbon offset utilization mechanism: methodology, the compatibility between local and national levels; connectivity among trading systems in various areas (quota, carbon offset)			
		The sustainability and quality of projects: the environmental, economic and social impact of different types of projects; radiation effect: does it produce the social effect that price serves as a lever to stimulate the eastern region to subsidize the western region, the rich to subsidize the poor?			
	4. Measurable, reportable and verifiable (MRV)	Is there third-party verification? How can the independence and accuracy of the third party (qualification management, social integrity) be guaranteed?			

TABLE 18.2 Overview of the non-governmental evaluation system (*cont.*)

Main evaluation item	Evaluation subitem	Indicator	Environmental impact	Social impact	Economic impact
		Is it complete and clear? Is a balance between scientific rigorousness and operability achieved? Is there MRV by industry and product?			
		Is there the risk of repeated calculation (e.g. inclusion of carbon offset projects, the electric power industry)?			
B. Mechanism execution	1. Performing unit	Is responsibility assumed independently and are the responsible party and the responsibility boundary specified? Is there cross responsibility or is the responsibility vague (e.g. subsidiary)?			
	2. Organizational setup	Are rights and responsibilities well-defined and is there a strong leadership? What about the coordination among relevant bodies involved and the capability of the personnel within bodies? In particular, with increasingly meticulous carbon trading, guidance is needed, rules need to be modified, expert competence is needed internally			
	3. Legal force	Is there a legal basis? What about validity? What about support at the national level?			

TABLE 18.2 Overview of the non-governmental evaluation system (*cont.*)

Main evaluation item	Evaluation subitem	Indicator	Environmental impact	Social impact	Economic impact
	4. Performance and punishment	Is there a compliance and punishment mechanism? What about a punishment standard?			
	5. Use of revenue	Does the system bring government revenue? If so, how can it be used?			
	6. Government intervention mechanism	How to intervene (directly or indirectly)? Who is in charge of it? What about the procedure and degree? Does it effectively play a role?			
C. Market performance	1. Trading volume and trading amount	Are the specific trading volume and trading amount disclosed?			
	2. Trader and platform	Is it easy to use? Is it safe? What about quantity and market segmentation?			
	3. Carbon price and its volatility	Are there floor prices, ceiling prices?			
	4. Carbon finance	Are there related products? Is there a secondary market?			
	5. Register, reporting system	Is it easy to use? Is it safe?			
	6. Effectiveness for enterprises	Is there an impact on enterprises' short-term, medium and long-term decision-making? Does it trigger enterprises' emission reduction action and the establishment of a carbon management process? What are the enterprises' future expectations?			

TABLE 18.2 Overview of the non-governmental evaluation system (*cont.*)

Main evaluation item	Evaluation subitem	Indicator	Environmental impact	Social impact	Economic impact
D. Information transparency and participation of stakeholders	1. Information disclosure to the general public	The degree, frequency and timeliness of information disclosure		▓	▓
	2. Information disclosure and interaction with enterprises	The degree, frequency and timeliness of disclosure; interaction with enterprises during policy making and after system operation has started		▓	▓
	3. Capacity building	Quantity, quality, depth and continuity		▓	▓
	4. Social supervision path	Does it exist? What about process, mechanism and effect? Note: the shaded area means that the indicator reflects the impact of this part		▓	▓

Note: The shadow part indicates that this index reflects the influence of this part.

SOURCE: GREENOVATION HUB: AN NGO REVIEW OF CHINA'S CARBON MARKET

CHAPTER 19

Perverse Incentives: Electricity Generation through Waste Incineration and Renewable Energy Subsidy

The Rock Environment & Energy Institute

In July 2014, the National Development and Reform Commission released the *Catalogue of Low-carbon Technologies Designated by the State as Promotional Priorities* (*Exposure Draft*), listing 34 low-carbon technologies. Against the realistic background of garbage siege and low carbon emission reduction pressure, the technology for electricity generation through waste incineration was also included in the list. Previously, the *Law of the People's Republic of China on Renewable Energy* and the renewable energy policies consisting of several normative documents issued by the National Development and Reform Commission included electricity generation through waste incineration as one category of biomass energy power generation in the scope of renewable energy electricity subsidy. However, the following issues deserve attention: whether China's technology for electricity generation through waste incineration is low-carbon, clean and sustainable, whether national subsidy for electricity generation through waste incineration is rational and efficient.

This chapter studies the relationship between China's renewable energy policy and electricity generation through waste incineration, the concept of renewable energy, the performance of the technology for electricity generation through waste incineration in carbon emissions, cleanliness and sustainability, and it analyzes, by enumerating relevant facts, the negative impact of renewable energy subsidy relating to electricity generation through waste incineration; it concludes that, generally, the waste is equal to neither biomass nor biomass waste; although waste contains lots of biomass, its rate of contribution to the generation of electricity is very low and generally lower than 1/3 due to high kitchen waste components, high moisture content and low thermal value. Therefore, it is very far-fetched to fully subsidize power generation through garbage incineration by merely regarding it as a project for the generation of electricity via renewable energy.

This chapter also holds that even if electricity generation through waste incineration partially comes from the energy generated by biomass waste burning, biomass energy itself involves uncertainty in sustainability from the perspective of energy sustainability, while at present, China lacks an elaborate green electric power certification system, which is not suitable to promoting a more efficient classification and recycling of biomass waste.

Moreover, in reality, the environmental pollution caused by waste incineration and the resulting human health risk have become indisputable facts in China. As a result, the technology is still far away from the requirement that renewable energy should be clean.

In response to social discussions about whether or not waste incineration is low-carbon, this chapter offers a reliable comparative data analysis and concludes that electricity generation through waste incineration is not a low-carbon energy utilization mode because: first, in comparison to other waste disposal modes, the carbon dioxide emission from per ton garbage in electricity generation through waste incineration ranks No.2, only second to anaerobic landfill, but it is 8 times that of power generation from anaerobic marsh gas. Second, compared with other energy power generation technologies, carbon dioxide emission per megawatt hour in electricity generation through waste incineration is the highest and reaches 2.72t; this figure is much higher than that of solar power generation and wind power generation; it is also higher than that of fossil energy power generation, including natural gas, oil, even coal power generation.

Although recently the Central Government introduced some policies designed to curb the situation in which waste incineration projects are carried out to fraudulently seek renewable energy subsidies by mixed burning of fossil energy, these policies fail to fundamentally resolve the contradiction between the subsidy for electricity generation through waste incineration and the support for electricity generation from renewable energy sources; this is because even if electricity generation through waste incineration does not entail mixed burning of fossil energy, its electricity subsidy mostly supports power generation from non-renewable energy sources, which apparently runs counter to the original purpose of the subsidy policy.

Overall, this chapter suggests that relevant government departments should cancel the electricity price subsidy for electricity generation through waste incineration as specified in the current renewable energy development policy—the portion of the current unified benchmark electricity price in excess of the benchmark on-grid price for desulfurization-related coal-fired units in various areas—and should carefully consider the policy system for including the technology for electricity generation through waste incineration in the *Catalogue of Low-carbon Technologies Designated by the State as Promotional Priorities*.

The correct substitute policy measures should cover two major aspects: First, change the electricity subsidy to payment of disposal charges, thus reform the system of waste disposal charge collection, put into practice the principle that whosoever generates waste should pay charges and whosoever generates more waste should pay more charges, so as to stimulate waste reduction, classification and recycling. Second, put the limited renewable energy electricity subsidy into cleaner and more sustainable waste management measures which are more conducive to reducing greenhouse gas emissions, provide various kinds of support including fund subsidies, for example, building the extended producer responsibility system, establishing a charge collection system able to reflect the real cost of waste disposal, taking part in the practice of waste classification, adopting the classification-based garbage recycling technology, especially pushing forward biochemical disposal and reuse of biomass waste.

Table 19.2 compares the current carbon dioxide emissions from waste and the carbon emission reduction potential under six different disposal modes in China. The result shows that waste incineration generates 0.815t carbon dioxide from per ton garbage; the net carbon dioxide emission after the alternative power grid carbon dioxide emission reduction from electricity generation is included, is 0.575t; it is the disposal method with the second highest carbon emission.

TABLE 19.1 Comparison of contribution rates of waste components in unit electric quantity

	Share of total weight	Electricity generation rates by waste type	Contribution toward total electricity generation from unsorted waste	Contribution generation toward 1 kilowatt-hour
Waste components	kg/kg garbage A	kWh/kg B	kWh/kg garbage C=A·B	% D=C/E
Kitchen waste	0.48	0.04	0.02	7.64
Paper	0.13	0.36	0.05	18.28
Plastics	0.12	0.96	0.12	46.60
Glass	0.03	0.00	0.00	0.00
Wood	0.01	0.36	0.00	1.20
Fabric	0.01	0.37	0.01	2.07
Others	0.22	0.28	0.06	24.21
Total	1.00		0.25 (E)	100.00

Note: Take the proportions of waste components at Luodai Garbage Incineration Power Plant, Chengdu City, Sichuan Province as an example, calculate the electricity generation contribution rates of components according to the rate of electricity generation by waste incineration of each waste component.

TABLE 19.2 Comparison of carbon emissions under different waste disposal modes

Ton gas/ton garbage	CH_4	CO_2	CO_2 equivalent total	Power grid CO_2 emission substitution a	Net CO_2 emission
	A	B	C=A·GWPb+B	D	E=C-D
Anaerobic marsh gas power generation c		0.256	0.256	0.186	0.07
Landfill+ marsh gas power generation	0.009	0.234	0.423	0.149	0.274
Power generation through incineration		0.815	0.815	0.240	0.575
Aerobic composting		0.334	0.334		0.334
Landfill+marsh gas burning	0.009	0.234	0.423		0.423
Anaerobic landfill	0.047	0.128	1.108		1.108

Note: GWPb refers to Global Warming Potential

APPENDIX I

Chronicle of Major Events in 2014

January 2: Corporate Environmental Credit Evaluation

On January 2, 2014, the Ministry of Environmental Protection (MEP), the National Development and Reform Commission (NDRC), the People's Bank of China (PBC), and the China Banking Regulatory Commission (CBRC) co-issued the *Environmental Credit Evaluation Measures for Companies (Trial)* in order to guide local authorities in assessing corporate environmental credits, and to urge companies to fulfill legal obligations and social responsibility for environmental protection, thus restraining them from losing environmental credit and punishing the non-compliant ones.

January 7: Air Pollution Prevention and Control

In order to implement the Air Pollution Prevention and Control Action Plan, the MEP signed the Letter of Responsibility for Air Pollution Prevention and Control Objectives with all the 31 provinces, autonomous regions, and municipalities directly under the Central Government) in the Chinese Mainland, clarifying air quality improvement objectives and key tasks for each region. In addition to the annual average PM2.5 concentration drop, this document includes the main tasks and measures specified in the aforementioned plan.

January 15: Energy Mix Adjustment

Wu Xinxiong, former Deputy Director of the NDRC and former Director of the National Energy Administration (NEA), disclosed that in 2014, the non-fossil energy consumption percentage would continue increasing to 10.7%, the non-fossil energy power generation percentage would reach 32.7%, the natural gas consumption percentage would rise to 61%, and the coal consumption percentage would fall to below 65%.

January 21: Ecosystem Protection

2014 was the 11th consecutive year when the Central Government's No. 1 document of the year emphasizes issues regarding agriculture as well as rural areas and people, but it was the first year when this document places ensured agricultural sustainability at the core, as it mentioned the necessity of balancing between high output/efficiency and sustainable use of resources.

January 23: Air Pollution Prevention and Control

In order to address severe smog, the NEA set up the Air Pollution Prevention and Control Work Office and issued the *Air Pollution Prevention and Control Work Program for the Energy Sector*.

February 5: Ecosystem Protection

The MEP issued the *National Red Lines for Ecosystem Protection—Technical Guide to Determining Ecological Function-specific Baselines (Trial)*. This was China's first ever technically guiding outline for defining red lines regarding ecosystem protection.

In 2014, the MEP finished defining red lines regarding ecosystem protection across China.

February 7: Air Pollution Prevention and Control

The NEA clarified the deadlines and responsibilities for key tasks in the area of air pollution prevention and control, for which it took five important measures.

February 14: Water Resources Management

Ten national authorities such as the Ministry of Water Resources (MWR) co-issued the *Implementation Plan for Implementing the Most Rigorous Assessment of Water Resources Management Systems* with clear stipulations relevant to assessors, procedures, contents, and scores. It signified the full implementation of the most stringent water resources management assessment and accountability measures in China.

February 13: Air Pollution Prevention and Control

Targets of total pollutant emission reduction in 2014 include: Sulfur dioxide, Chemical Oxygen Demand (COD), and ammoniacal nitrogen (NH3-N) each by 2% and NOx emissions by 5%. In 2014, 1,473 projects should be completed in all Chinese regions; denitrification at power plants should increase by another 130 million kilowatts; denitrification for iron and steel sintering equipment should increase by 15,000 square meters; three million heavily-polluting cars should be scrapped.

February 25: Air Pollution

For many Shijiazhuang residents, an air pollution index of more than 300 seems to have been not uncommon. However, Li Guixin, a resident in Xinhua District, Shijiazhuang

City made an unusual move. He filed an administrative petition to the People's Court of Yuhua District, Shijiazhuang City.

March 3: Air Pollution Control

The Beijing Municipal Environmental Protection Bureau (BJEPB) organized a symposium on joint prevention and control of air pollution in the Beijing-Tianjin-Hebei area and surrounding regions. It proposed to establish a regional heavy-pollution warning consultation and emergency response mechanism so that relevant authorities and organizations could discuss and take action in a unified manner for joint response if regional heavy air pollution is predicted.

March 7: Action against Violation of Environmental Law

While the MEP issued the *Urgent Notice on Special Environmental Inspections*, many local governments immediately launched various inspections on the violation of environmental law and issued notices ordering violators to pay a hefty fine. A number of experts told reporters that companies may enter the era of high-cost environmental violations due to the possibility of paying a huge fine.

March 14: Drinking Water Quality

The MEP released the results of its first ever national large-scale survey on drinking water quality. The results showed that there were 250 million people living in areas near heavily-polluting enterprises and trunk roads, and that 280 million people were using unsafe drinking water.

March 14: Environmental Proposals

The Second Session of the 12th Chinese People's Political Consultative Conference (CPPCC) National Committee ended on March 14, 2014. During the meeting, members of this committee as well as participating organizations and subcommittees of the CPPCC all offered proposals. As of the 7th day of March at 14:00, a total of 5,875 proposals were submitted. Of them, a total of 596 proposals were about strengthening pollution prevention and ecosystem rehabilitation; 204 others, action against environmental pollution.

March 17: Water Resources Protection

The NDRC took the lead in formulating the *Guiding Opinions on Relying on the Yangtze River to Build New Economic Support Zones in China*, signifying the implementation of the most stringent water resources management system throughout the Yangtze River Basin. By clarifying red lines for the development and utilization of water resources from the Yangtze River, this document aims to strictly control the over-exploitation of water resources.

March 17: Wetland Protection

The State Forestry Administration (SFA) issued the *Warning Plan for the Change of Ecological Characteristics of Internationally Important Wetlands (Trial)*. China will implement a three-level (i.e., yellow, orange, and red) warning system on changes in ecological characteristics of internationally important wetlands.

March 25: Climate Change

Researchers at the University of Leeds discovered through studies on climate models and food production that from 2030 on, global maize, wheat, and rice production will begin to decline with climate change, with negative impact significantly earlier than expected.

March 28: Air Pollution

The World Health Organization (WHO) stated that air pollution had become the world's primary source of environmental health risks. In 2012, for example, as many as seven million people died from air pollution.

April 3: Carbon Emissions Trading

The first pilot carbon emissions trading program in Central China was officially launched at Wuhan Optics Valley United Property Rights Exchange (OVUPRE) on April 2, 2014. Earlier, Hubei Province had successfully conducted China's first ever bid-based transfer of government-reserved carbon emission allowances. Hubei was the 6th pilot province that initiated carbon emissions trading following Shenzhen, Shanghai, Beijing, Guangdong Province, and Tianjin. At present, a total of 138 Hubei-based enterprises in 12 industries such as electricity, thermal power and iron & steel have been included in carbon emission allowance management. In 2014, carbon emission allowances in Hubei totaled 324 million metric tons, second only to Guangdong.

April 11: Drinking Water Pollution

Beginning from the afternoon of April 11, 2014, Lanzhou Veolia Water Group as the only water supplier in Lanzhou City, Gansu Province, was found to have severely excessive benzene in samples of both drinking water delivered to local residents and water from its pipeline. The detected benzene value was 200 μg/L at 14:00, in particular, which was 20 times the national limit. Excessive benzene was also found in tap water at individual resident apartments in the city. The local government stated that tap water was not suitable for drinking within 24 hours. An investigation on pollution sources is still in progress. The Yellow River water is not polluted, according to the local government.

April 17: Soil Pollution

The MEP and the Ministry of Land and Resources (MLR) co-issued the *National Soil Pollution Survey Report* in order to publish the results of this eight-year survey. According to this report, the overall percentage of non-compliant spots is 16.1%; the percentage of non-compliant arable land spots, 19.4%; the percentage of non-compliant forest spots, 10.0%; the percentage of non-compliant grassland spots, 10.4%. Soil pollution is heavier in the south than in the north.

April 22: Groundwater Quality

The MLR released *China Land and Resources Report 2013* on April 22, 2014, showing that of monitoring results from 4,778 groundwater quality monitoring sites in 203 prefecture-level cities across China, nearly 60% was "Poor" or "Very Poor."

April 22: Marine Protection

The State Oceanic Administration (SOA) released the *Notice of the State Oceanic Administration on the Approval of Establishing Eleven National Special Marine Protection Areas (Ocean Parks) Including Panjin Yuanyanggou National Ocean Park*, adding 11 national special marine protection areas (ocean parks). At this point, China already has 56 such areas with a total area of 69,000 square kilometers, including 30 ocean parks.

April 23: Drinking Water Pollution

In the afternoon on April 23, 2014, excessive NH3-N concentration was found in the Wuhan section of the Hanjiang River, resulting in non-compliant NH3-N concentration of water supplied by Wuhan Baihezui Water Plant and Yushidun Water Plant in Dongxihu District, Wuhan City, both of which take water from this section. Both of them stopped production on the day at 16:34 and 19:30 respectively. Contaminated tap water did not enter the pipe network, but water pressure dropped in some areas. The city stopped water supply in an area of 260 square kilometers, affecting more than 300,000 residents and hundreds of food processing companies.

April 24: New Environmental Law Enacted

The Eighth Session of the Standing Committee of the 12th National People's Congress (NPC) approved the *Environmental Protection Law (Amendment)*. The draft received four deliberations within 20 months before it was finalized. With increased government and corporate responsibilities and penalties in various aspects, It was described by some experts as "the strictest environmental protection law ever."

May 2: Environmental Liability Insurance

In China, the cost of an environmental pollution incident tends to be ultimately paid by the government. But this will change in Anshan City, Liaoning Province. From May onward, Anshan Municipal Government will conduct a pilot environmental liability insurance program among companies with high environmental risks in an attempt to protect the environment in a market-oriented manner.

May 9: Water Pollution Incident

On the morning of May 9, 2014, water supply was stopped in Jingjiang City, Jiangsu Province, due to abnormal water quality at the local drinking water source, affecting nearly 700,000 local people at work and in life and causing a rush for clean drinking water.

May 21: Clean Energy

On May 21, the NDRC announced the first batch of projects, 80 in all, for which private investment is encouraged, in areas such as infrastructure. Of these, 36 are in the field of clean energy, including 30 distributed photovoltaic generation demonstration area projects in the field of solar power alone.

May 28: Draft of Environmental Tax Law for Deliberation

The Legislative Affairs Office of the State Council (LAO) solicited the opinions of relevant departments on the draft of the environmental tax law for deliberation. Next, it will submit this draft to the State Council for consideration and then, with the latter's approval, submit it to the NPC Standing Committee for deliberation.

May 25: Ecological Degradation

China is now facing very serious environmental problems in agriculture, as exemplified by severe degradation of its agricultural ecosystem. Wetland area has decreased by 3.4 million hectares in the past decade; the total salinized land area has reached 12 million hectares; 90% of the natural grasslands have been degrading in different degrees; the average overload rate of northern grasslands is 36%. About 66,667 hectares of cultivated land is lost per year on average due to soil erosion.

June 4: Acid Rain

Chinese cities are in a tough situation in terms of ambient air quality. Acid rain mostly occurs along the Yangtze River and areas south of its middle and lower reaches. Acid rain-affected areas represent about 10.6% of China's total land area.

June 4: Seawater Quality

In 2013, the overall seawater quality of coastal waters in China was average. The proportion of Levels I and II seawater is 66.4%; that of Levels III and IV seawater, 15.0%; that of sub-Level IV seawater, 18.6%. Of China's four seas, both the Yellow and South China Seas feature good coastal seawater quality; the Bohai Sea, normal; the East China Sea, extremely poor.

June 5: Air Quality

The MEP released the *2013 Report on the State of the Environment of China*, showing that in 2013, Haikou, Zhoushan and Lhasa were the only ones of 74 key Chinese cities which met the new Ambient Air Quality Standards, namely, merely 4.1% of them were compliant. This year saw 35.9 smoggy days on average in China, or 18.3 days more than in the previous year, setting a new high since 1961.

June 9: Environmental Information Transparency

An environmental non-governmental organization (NGO) announced the rankings of 120 Chinese cities by environmental information transparency index. Improvements include significant progress in online real-time release of monitoring results, and systematic release of daily regulatory information in more cities. On the other side, there remain an urgent need to establish a corporate emission data disclosure system; it is still necessary to improve the systems for public access to environmental assessment results and participation in such assessment.

June 17: Nineteen Companies Fined by MEP

The MEP announced the highest fine ever, ordering 19 companies to pay a total of 410-million-yuan worth of desulfurization-specific electricity price or additional pollutant emission charge for severe problems with their desulfurization facilities. Power companies accounted for the majority of these companies, as subsidies of China's Top 5 power groups, i.e., China Huaneng, China Guodian, China Huadian, China Datang and SPIC (State Power Investment Corporation), were on the list.

June 17: Environmental Assessors Criticized for Report

The MEP publicly criticized three environmental assessors such as China Waterborne Transport Research Institute (WTI) for the *Environmental Impact Report on Jinzhou Port's Coal Terminal Project (Changes)*. The MEP did not think that the report's conclusions on the post-change impact on the atmospheric environment was credible.

July 3: Environment- & Resources-specific Trial Court

In order to respond actively to the latest public expectations for environment-and resources-relevant justice and provide strong judicial support for building an eco-civilization, the Supreme People's Court decided to set up an environment- and resources-specific trial court.

July 10: Coal Causes High External Environmental Costs, Incl. Over 300 Billion Yuan Health Losses

Coal has caused huge external environmental costs in China. The total cost in 2010 was 555.54 billion yuan, of which air pollution-caused human health losses, especially those of workers in mining areas, was the greatest at 305.1 billion yuan, or 55% of the total external environmental cost, according to a report titled "Coal's External Environment Costs: Accounting and Internalization Scheme."

June 10: Damages to Marine Ecosystems

The first United Nations Environment Assembly (UNEA) was opened in Nairobi. Two reports released on the same day pointed out that a large amount of plastic waste in the oceans constitute a growing threat to the survival of marine life; it is conservatively estimated that the resulting annual loss of marine ecosystems is as high as US$13 billion.

July 11: List of Air Polluters

The MEP's official website published the *Draft of Technical Guidelines for Making a List of Air Polluters (Trial) for Opinions* involving five types of polluters such as sources of dust in cities, motor vehicles on roads, and biomass combustion sources. This document will offer technical support for subsequent listing of air pollution sources.

June 16: Polluters Fined

Ten power companies were ordered by local pricing authorities to pay 519 million yuan in fines for problems such as abnormal operations of desulfurization facilities despite that they had received subsidies for desulfurization-specific electricity price. Subsidiaries of state-owned energy companies such as China Guodian and China Huadian were involved.

July 23: Environmental Cases Reported by Public

The MEP released information on environmental cases reported by the public via its 12369 Environmental Hotline in the 1st half of 2014. The information showed that 696 cases were reported in that period, and that about 80% of them did involve illegality.

August 1: Air Pollution Control

Beijing Municipal Government initiated the compulsory use of Beijing Standard Coal, or low-sulfur coal. This move will contribute to further improvements in local air quality.

August 4: Beijing Issues Strictest No-Burning Zone Scheme, Banning Use of Some Heavily-polluting Fuels

Beijing Municipal Government officially released the *Beijing No-Burning Zone Delineation Scheme for Highly-polluting Fuels (Trial)*. More than ten types of highly polluting fuels such as raw (or bulk) coal, pulverized coal, fuel oil, petroleum coke, combustible waste, and biomass fuels for direct combustion are prohibited from being used in specific areas.

August 4: Wetland Protection

In order to promote wetland protection/restoration and the building of an eco-civilization, the Central Government began increasing forestry subsidies in 2014, and supported the launching of projects such as the transforming of agricultural land into wetland, a pilot compensation program for ecological benefits from wetland, and wetland conservation incentives.

August 5: Soil Remediation

The MEP organized the drafting of the *Soil Environmental Protection Law*. A draft has been made, and relevant departments and local authorities have been solicited for their opinions on it. Next, the MEP will further strengthen field studies on soil legislation, actively improve the draft law, and systematically advance soil legislation.

August 22: Pollution Control

In order to guide local governments in strengthening environmental regulation on land transfer, redevelopment/reuse, completion/acceptance and other major activities specific to contaminated sites, and facilitate the resolution of key regulatory problems, the MEP issued the *Circular on Launching Pilot Environmental Regulation for Contaminated Sites*.

September 2: Heavy Metal Pollution Control

The second inter-ministerial meeting on the prevention and control of heavy metal pollution was held in Beijing on September 2, 2014. In the past three years, the Central Government has allocated a total of 11.6 billion yuan to control heavy metal pollution, attracting more than 30 billion yuan of investment by local governments and companies. As a result, over five million metric tons of copper, lead, and zinc smelting capacity was cut.

September 6: Environmental Pollution

Sewage ponds appeared in parts of the central Tengger Desert in the Inner Mongolia Autonomous Region. Local companies discharged untreated wastewater into these ponds, where it was left evaporating naturally. The sticky sediment was then shoveled out with a forklift and buried directly in the desert.

September 13: Waste Incineration

Some people in Boluo County, Huizhou City, Guangdong Province, drew public attention as they took to the street to protest a proposed waste incineration plant. This waste incineration plant was one of a series of projects in Huizhou Ecological Park.

September 22: Species Protection

In September 2014, a number of research institutes including the Yangtze River Fisheries Research Institute (YFI) confirmed that in 2013, Chinese sturgeon did not breed or lay eggs at the only natural spawning ground down Gezhouba Dam. The wild population of Chinese sturgeon has been shrinking since this dam was built 32 years ago.

September 23: Climate Negotiations

The Climate Summit 2014 was closed at the UN headquarters in New York in the evening on the 23rd day of September. This summit is the largest international conference on climate change, and the meeting of this year is the most important of its type before the 2015 United Nations Climate Change Conference takes place in Paris.

October 9: Environmental Pollution

There are currently more than 100,000 mines in China, which have damaged a total of 3.868 million hectares of land, affected 5.38 million hectares of underground aquifers, and caused a cumulative total of 40 billion metric tons of solid waste, plus over 4.7 billion cubic meters of discharged wastewater per year, according to a statement at China Mining Green Development Exchange Conference and Economic Transformation Summit Forum 2014.

October 10: Energy Conservation and Emission Reduction

In order to promote technological upgrading and modification at coal power plants for energy conservation and emission reduction, the NEA will formulate a scientifically sound electric power development plan, determine a reasonable annual capacity limit of new coal-fired power plant projects, and guide local optimization of such projects. The aforementioned limit will be strictly controlled in areas with an unreasonably small number of hours of coal power utilization (below 4,500 hours).

October 11: Ecosystem Protection

The Forestry Department of Sichuan Province (FDSC) issued the *Outline of the Plan of Promoting Eco-friendly Forestry Development in Sichuan Province (2014–2020)*, defining the province's first ever ecological red lines for woodland/forests, wetlands, sand vegetation, and species. By 2020, the local woodland/forest/wetland areas will be no less than 23.6, 18, and 1.67 million hectares respectively; a healthy group of ecosystems will appear in the Sichuan section of the Yangtze River's upper reaches.

October 17: CO_2 Emission Drop

On the 14th day of October, the NDRC and other relevant authorities officially launched field assessment on the performance of responsibility for achieving CO_2 emission reduction targets per unit of GDP, as required by the *Notice of the National Development and Reform Commission on Issuing the Measures for Assessing the Performance of Responsibility for Achieving CO_2 Emission Reduction Targets per Unit of GDP*.

October 20: Air Pollution Prevention and Control

In order to implement the Air Pollution Prevention and Control Action Plan, urge local governments to enhance environmental regulation on sources of air pollution, improve ambient air quality, and ensure winter air quality safety, the MEP launched supervision on air pollution prevention and control in the winter of 2014.

November 4: Third-party Environmental Pollution Management

The executive meeting of the State Council required the implementation of third-party environmental pollution management and the promotion of government procurement of private environmental monitoring services. Polluting companies were required to sign contracts with environmental service providers in order to achieve the goal of compliant emissions/discharges by purchasing pollutant emission reduction services, and to jointly monitor the effectiveness of third-party environmental pollution management with relevant environmental regulators.

November 5: Air Pollution Control

During the APEC meeting, the Beijing-Tianjin-Hebei area and the six provinces around it launched the greatest efforts for temporary emission reduction after the Beijing Olympic Games. The MEP sent 16 inspection teams to initiate large-scale inspections in conjunction with municipal and provincial environmental authorities. Local governments and environmental authorities also set up various supervision teams for daily inspections.

November 7: Soil Remediation

The *Soil Pollution Prevention and Control Law (Proposal)* drafted by the MEP was submitted to the NPC Environment and Resources Conservation Committee (ERCC) at the end of 2014. It will be handed over to the LAO after the ERCC conducted coordination according to all stakeholders' opinions.

November 12: US-China Joint Announcement on Climate Change

According to the US-China Joint Announcement on Climate Change, The United States intends to achieve an economy-wide target of reducing its emissions by 26%–28%

below its 2005 level in 2025; China intends to achieve the peaking of CO2 emissions around 2030 and to make best efforts to peak early and intends to increase the share of non-fossil fuels in primary energy consumption to around 20% by 2030.

November 12: APEC Blue

China National Environmental Monitoring Center (CNEMC) data showed that the air quality in Beijing was good during November 1–12, 2014, except for the 4th day of the month, when the air quality index (AQI) indicated slight pollution. As the APEC meeting took place from the 7th to 12th days of the month, such a sky color has since been referred to as "APEC blue."

November 14: Soil Pollution Control

A series of legislative jobs specific to soil pollution control were coming to an end in 2014. The *Soil Pollution Prevention and Control Law (Proposal)* had been finalized and submitted to the NPC; the Soil Pollution Prevention and Control Action Plan will soon be submitted to the State Council for deliberation and is expected to be issued at the end of 2014 or the beginning of 2015.

November 17: Ecosystems along Yangtze River

Building a green ecological corridor is one of the key tasks, according to the *Guidelines for Relying on the Golden Waterway to Promote the Development of the Yangtze River Economic Belt* issued by the State Council.

December 1: Climate Change

The 20th yearly session of the Conference of the Parties (COP 20) to the 1992 United Nations Framework Convention on Climate Change (UNFCCC) was held in Lima, Peru. The Lima COP was to hold discussions around three key issues including agreement on the elements of a formal, draft negotiating text, agreement on the framework and information required for so-called Intended Nationally Determined Contributions, and agreement on actions countries will voluntarily take to enhance ambition in the pre-2020 period, according to the authorization of the Warsaw COP.

December 6: Environmental Pollution

Suntown Technology Group began building Changde Industrial Park (i.e., Chuangyuan Aluminum), which is 163.3 hectares in area, in Taoyuan County, Changde City, Hunan Province in 2001. Heavily-polluting, energy-hungry projects such as rough aluminum processing have since been launched one after another, causing serious local environmental pollution. Several villagers have died of cancer. Many farmers have been forced to leave their hometowns and become environmental migrants.

December 8: Soil Contamination

China's total non-compliance rate of soil contamination is 16.1%; specifically, the proportions of spots with slight, mild, moderate, and severe pollution are 12.1%, 2.3%, 1.5% and 1.1% respectively; Levels I and II soil suitable for agricultural cultivation in China accounted for 87.9%; soil with potential ecological risks accounted for 12.1%, of which moderately and heavily polluted soil accounted for about three percentage points, according to the *National Soil Pollution Survey Report* published by the MEP and the MLR.

December 18: Heavy Metal Pollution

The MEP released 2013 annual assessment results under the Twelfth Five-Year Plan for Comprehensive Prevention and Control of Heavy Metal Pollution: By the end of 2013, the total discharge of five key heavy metal pollutants (i.e., lead, mercury, cadmium, chromium, and arsenic as a metalloid) in China dropped by 10.5% from the previous year.

December 18: Arable Land Quality

On December 17, 2014, the Ministry of Agriculture (MOA) issued the *Report on the Quality of Cultivated Land in China*, classifying arable land for the first time in China. Levels I, II and III arable land is 33.2 million hectares in area, accounting for 27.3% of China's total cultivated land area; Levels IV, V and VI cultivated land is 54.5 million hectares in area, accounting for 44.8% of the total cultivated land area; Levels VII through X arable land is 34 million hectares, accounting for 27.9% of the total cultivated land area.

December 19: Organic Pollution Source Identification

Schemes of paying for volatile organic waste discharge will be studied and formulated, and will first be implemented in the petrochemical industry, according to the *Comprehensive Plan for the Control of Volatile Organic Compounds in the Petrochemical Industry* released by the MEP.

December 22: Air Pollution Prevention and Control Law

The Twelfth Session of the Standing Committee of the 12th NPC heard a statement from the State Council concerning the proposal for deliberation on the *Air Pollution Prevention and Control Law of the People's Republic of China (Draft Amendment)*, which contains new provisions for response to heavy pollution, administrative penalties, total discharge control and emission permits, air pollution prevention and control in key fields and regions, etc.

December 25: New Environmental Law

The new *Environmental Protection Law* will formally take effect on January 1, 2015. The new law not only defines a variety of penalties, such as fine calculation by day and immediate production limitation or suspension for excessive pollutant discharge, but also allows the detention of relevant directors and people directly responsible for pollution in four types of situations such as unauthorized construction and unlicensed pollutant discharge. Besides, it authorizes environmental law enforcement officials to close down relevant facilities and/or impound relevant items.

APPENDIX II

Green Book of Environment Evaluation of China's Environmental Performance in 2014

Theme 1: Total Pollutant Discharge[1]

Condition

In 2013, China meets the annual discharge and emission reduction targets for major pollutants of chemical oxygen demand (COD), ammonia nitrogen, sulfur dioxide and nitrogen oxide

Major Pollutants in Wastewater

In 2013, total COD discharge is 23.527 million tons, down 2.9% from the previous year. Discharge of ammonia nitrogen is 2.457 million tons, down 3.1% from the previous year.

TABLE II.1 Discharge of major pollutants in wastewater nationwide in 2013

Pollutants	Chemical oxygen demand (Million tons)	Ammonia nitrogen (Million tons)
Total discharge	23.527	2.457
Industrial source	3.195	0.246
Domestic source	8.898	1.414
Agricultural source	11.257	0.779
Centralized facilities	0.177	0.018

Major Pollutants in Exhaust Gas

In 2013, total emission of sulfur dioxide is 20.439 million tons, down 3.5% from the previous year. Total emission of nitrogen oxide is 22.273 million tons, down 4.7% from the previous year.

---

1 Data source: China Environmental Bulletin 2014, issued by the Ministry of Environmental Protection of China, June 6, 2015.

Solid Waste

In 2013, total generation of industrial solid waste nationwide is 3.277019 billion tons, the combined utilization of industrial solid waste (including the utilization of stored waste in previous years) is 2.059163 billion tons, and the combined utilization rate is 62.3%.

TABLE II.2   Emission of major pollutants in exhaust gas nationwide in 2013

Pollutants	Sulfur dioxide (Million tons)	Nitrogen oxide (Million tons)
Total discharge	20.439	22.273
Industrial source	18.352	15.457
Domestic source	2.085	0.407
Vehicle	—	6.405
Centralized facilities	0.002	0.004

TABLE II.3   Generation and utilization of industrial solid waste nationwide in 2013

Generation (Million tons)	Combined utilization (Million tons)	Stored quantity (Million tons)	Treated quantity (Million tons)
3,277.019	2,059.163	426.342	829.695

Theme II: Water

Rivers

Among the state-level monitored cross-sections of the 10 largest river basins—namely Yangtze River, Yellow River, Pearl River, Songhua River, Huaihe River, Haihe River, Liaohe River, rivers in Zhejiang and Fujian provinces, rivers in northwest China, and rivers in southwest China—water quality in Category -III, in Category IV- and worse than Category V account for 71.7%, 19.3% and 9.0% respectively. Water quality has no significant changes compared with the year 2012. Major pollutants are COD, permanganate index and biochemical oxygen demand after 5 days (BOD_5).

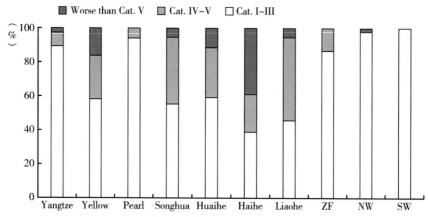

FIGURE II.1   Comparison of China's 10 largest river basins in 2013 by categories of water quality

TABLE II.4   Comparison of China's 10 largest river basins in 2013 by categories of water quality

	Yangtze	Yellow	Pearl	Songhua	Huaihe	Haihe	Liaohe	ZF*	NW**	SW***
Cat. I–III	89.4%	58.1%	94.4%	55.7%	59.6%	39.1%	45.5%	86.7%	98.0%	100.0%
Cat. IV–V	7.5%	25.8%	0.0%	38.6%	28.7%	21.8%	49.1%	13.3%	0.0%	0.0%
Worse than Cat. V	3.1%	16.1%	5.6%	5.7%	11.7%	39.1%	5.4%	0.0%	2.0%	0.0%

* ZF: Rivers in Zhejiang and Fujian provinces
** NW: Rivers in northwest China
*** SW: Rivers in southwest China

Lakes and Reservoirs

In 2013, the shares of the state-level monitored lakes and reservoirs with good, slightly polluted, moderately polluted and severely polluted water quality are 60.7%, 26.2%, 1.6% and 11.5% respectively. Compared with the year 2012, major pollutants are total phosphorus, COD and permanganate index.

TABLE II.5 Water quality of major lakes and reservoirs in 2013

Type of Lake/Reservoir	High quality	Good quality	Slightly polluted	Moderately polluted	Severely polluted
Three lakes*	0	0	2	0	1
Major lakes	5	9	10	1	6
Major reservoirs	12	11	4	0	0
Total	17	20	16	1	7

* Three lakes refer to Taihu Lake, Dianchi Lake and Chaohu Lake

Lakes and reservoirs that are eutrophic, mesotrophic and oligotrophic account for 27.8%, 57.4% and 14.8% respectively.

Groundwater

There are 4,778 monitoring points of groundwater environmental quality in 2013, among them 800 are state-level. Of these monitoring points, 10.4% has high water quality, 26.9% good water quality, 3.1% moderate water quality, 43.9% poor water quality, and 15.7% bad water quality. Major indexes exceeding the standards include total hardness, iron, manganese, total dissolved solids, nitrite, nitrate, ammonia nitrogen, sulfate, fluoride, chloride, etc.

TABLE II.6 Water quality of state-level monitored lakes and reservoirs during 2004–2013

Year	Cat. I, II, III	Cat. IV, V	Worse than Cat. V
2004	26.0%	37.0%	37.0%
2005	28.0%	29.0%	43.0%
2006	29.0%	23.0%	48.0%
2007	49.9%	26.5%	23.6%
2008	21.4%	39.3%	39.3%
2009	23.1%	42.3%	34.6%
2010	23.0%	38.5%	38.5%
2011	42.3%	50.0%	7.7%
2012	61.3%	27.4%	11.3%
2013	60.7%	27.8%	11.5%

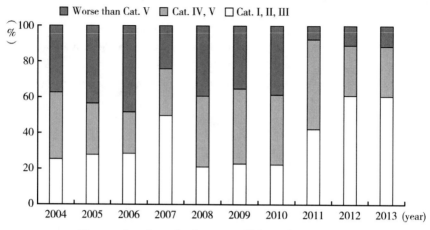

FIGURE II.2 Water quality of state-level monitored lakes and reservoirs during 2004–2013

Compared to the previous year, 4,196 points have continuous water quality monitoring data in 185 cities, and the water quality in general is stable. Among them, water quality in 15.4% of the points has improved, 66.6% remained stable and 18.0% deteriorated.

Key Water and Hydro Projects

Reservoir area of the Three Gorges Project has good water quality, and the water quality of all the three state-level monitored cross-sections is in Category III. Cross-sections with standard-exceeding total nitrogen and total phosphorus account for 90.7% and 77.9% respectively in the first-level tributaries. The comprehensive trophic state index of the tributary water bodies ranges between 28.8 and 73.0, and the share of the monitored cross-sections with eutrophic water quality is 26.6%. Dominant species in the tributary water bodies are bacillariophyta (cyclotella), cyanophyta (oscillatoria and microcystis), pyrroptata (peridinium) and cryptophyta (cryptophyta).

Water quality of the **South-to-North Water Diversion Project (East Route)** at the three cross-sections of Yangtze River intake, Jiajiang, and Sanjiangying is in Category III. Sections of the main line of the project (namely, the historical Beijing-Hangzhou Grand Canal), including the Inner Canal, Baoying Canal, Suqian Canal, Lunan Canal, Hanzhuang Canal and Liangji Canal, also have good water quality. Compared with the year 2012, water quality of Liangji Canal has improved, and no significant changes can be found for the other sections.

Hongze Lake is moderately polluted and lightly eutrophic with major pollutant of total phosphorus. Luoma Lake, Nansi Lake and Dongping Lake have good water quality and are mesotrophic. Yihe River (that flows into Luoma Lake) also has good water quality. Among the 11 rivers that flow into Nansi Lake, the Zhuzhao New River is lightly polluted with major pollutants of COD and oil, and the other rivers have good water quality. Dawen River (that flows into Dongping Lake) also has good water quality.

The Taocha cross-section of the intake of the **South-to-North Water Diversion Project (Middle Route)**, has its water quality in Category II. The Danjiangkou Reservoir has high water quality and is mesotrophic. All the 9 rivers flowing into the Danjiangkou Reservoir have high or good water quality. Compared with the year 2012, water quality of Tianhe River, Guanshan River and Laoguan River has gone down, while no significant changes are observed in the other rivers.

Seas
Condition

In 2013, the water environment of the Chinese sea areas is in general good, and the offshore sea area areas have moderate water quality.

Overall Sea Areas

In 2013, water quality of sea areas under the jurisdiction of People's Republic of China is in general good. Sea areas in Category I of seawater quality standard account for 95% of the total sea areas under Chinese jurisdiction.

Coastal Waters

In 2013, water quality of China's coastal waters is in general moderate. Calculated according to the monitoring points, seawater quality of Category I and Category II accounts for 66.4%, down 3.0 percentage points from the previous year; Category III and Category IV account for 15.0%, up 3.0 percentage points; and sea points with water quality worse than Category IV account for 18.6%, unchanged compared with the previous year. Exceeding concentration of major pollutants are inorganic nitrogen and labile phosphate.

Coastal waters of the Bohai Sea have moderate water quality. Sea points with water quality in Category I and II accounts for 63.2%, down 4.1 percentage points from the previous year; sea points in Category III and IV take up a share of 30.7%, up 10.2 percentage points; and those worse than Category IV have a share of 6.1%, down 6.1 percentage points. Major pollutants are inorganic nitrogen, lead and nickel.

Coastal waters of the Yellow Sea have good water quality. The share for sea points in Category I and II is 85.2%, down 1.8 percentage points from the previous year, while that in Category III and IV is 14.8%, up 1.8 percentage points. There are no sea points worse than Category IV, the same as the previous year. Major pollutants are inorganic nitrogen and oil.

Coastal waters of the East China Sea have bad water quality. The share for sea points in Category I and II is 30.5%, down 7.4 percentage points from the previous year; that in Category III and IV is 20.0%, up 4.2 percentage points; and that worse than

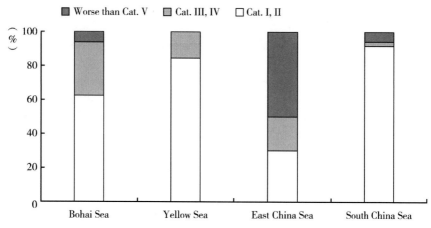

FIGURE II.3 Seawater quality of different coastal waters in 2013

Category IV is 49.5%, up 3.2 percentage points. Major pollutants are inorganic nitrogen, labile phosphate and biochemical oxygen demand (BOD).

Coastal waters of the South China Sea have good water quality. The share for sea points in Category I and II is 91.3%, up 1.0 percentage point; that in Category III and IV is 2.9%, down 1.0 percentage points; and that worse than Category IV is 5.8%, the same as the previous year. Major pollutants are inorganic nitrogen, labile phosphate and pH value.

Key Gulfs and Bays

Among the 9 key gulfs and bays, the Beibu Gulf has high water quality, the Yellow River estuary has good water quality, the Liaodong Bay, Bohai Bay and Jiaozhou Bay have poor water quality, while the Yangtze River estuary, Hangzhou Bay, Minjiang River estuary and Pearl River estuary have bad water quality. Compared with the year 2012, water quality has improved in Beibu Gulf and Bohai Bay, worsen in the estuaries of both Yellow River and Minjiang River, and the other gulfs and bays maintain a stable water quality.

Discharge of Land-Sourced Pollutants into the Sea Areas

In 2013, discharge of wastewater from the 423 monitored industrial, domestic and mixed sources, each of which has a daily discharge volume larger than 100 cubic meters, is 6.384 billion tons in total, and specifically COD 221,000 tons, oil 1,636 tons, ammonia nitrogen 16,900 tons, total phosphorus 2,841 tons, mercury 213 kilograms, hexavalent chromium 1,908 kilograms, lead 7,681 kilograms and cadmium 392 kilograms.

TABLE 11.7 Total discharge of pollutants into the 4 largest seas in 2013 (in tons)

Sea area	Wastewater	COD	Oil	Ammonia nitrogen	Total phosphorus
Bohai Sea	206 million	12,000	36.2	2,000	180.4
Yellow Sea	1,104 million	55,000	235.8	4,000	662.0
East China Sea	3,745 million	119,000	861.6	8,000	1,046.9
South China Sea	1,329 million	35,000	501.9	4,000	951.8

Theme III: Urban Environment

Air Quality

74 cities monitored with new standard in first phase: In 2013, totally 74 cities in key areas such as the Beijing-Tianjin-Hebei region, the Yangtze River delta and the Pearl River delta, as well as provincial capitals and cities with independent budgetary status, are monitored with the new standard. According to the Ambient Air Quality Standard (GB3095–2012), values are assessed of the annual average of SO_2, NO_2, PM_{10} and $PM_{2.5}$, the daily average of CO and the daily maximum eight-hour average of O_3. Among the 74 cities, only Haikou, Zhoushan and Lhasa have their air quality up to the standard, accounting for 4.1%, and 95.9% of the cities exceed the air quality standard. The top 10 cities with relatively better air quality are Haikou, Zhoushan, Lhasa, Fuzhou, Huizhou, Zhuhai, Shenzhen, Xiamen, Lishui and Guiyang, while the worst 10 include Xingtai, Shijiazhuang, Handan, Tangshan, Baoding, Jinan, Hengshui, Xi'an, Langfang and Zhengzhou.

The average percentage of the number of days out of a year on which air quality in these 74 cities met safety standards is 60.5%, and the average percentage of days on which air quality failed to meet those standards was 39.5%. Among them, 10 cities met safety standards is between 80% and 100%, 47 cities are between 50% and 80%, and 17 cities are below 50%.

Three key areas: In 2013, all cities in the Beijing-Tianjin-Hebei region and the Pearl River delta failed to meet the safety standards, and only Zhoushan in the Yangtze River delta meets the safety standards of all the six pollutants.

In 2013, the average percentage of days on which air quality in 13 cities at or above the prefecture-level in the Beijing-Tianjin-Hebei region met safety standards ranges from 10.4% to 79.2%, and the average is 37.5%. Of the total number of days that failed to meet safety standards, 20.7% are severely polluted. In 10 cities, the average

TABLE II.8 Numbers of cities in key areas meeting the safety standards of all pollutants

Area	Total	SO$_2$	NO$_2$	PM$_{10}$	CO	O$_3$	PM$_{2.5}$	Meeting all standards
Beijing-Tianjin-Hebei	13	7	3	0	6	8	0	0
Yangtze River delta	25	25	10	2	25	21	1	1
Pearl River delta	9	9	5	5	9	4	0	0

percentage of days on which air quality failed to meet standards accounts for less than 50% of all days in the year. Of the total number of days that failed to meet safety standards in the Beijing-Tianjin-Hebei region, more days have PM$_{2.5}$ as the major pollutant, taking a share of 66.6%. Days with PM$_{10}$ and O$_3$ as the major pollutant accounts for 25.2% and 7.6% respectively.

In 2013, the average percentage of days on which air quality in 25 cities at or above the prefecture-level in the Yangtze River delta met safety standards ranges from 52.7% to 89.6%, and the average is 64.2%. Of total number of days that failed to meet safety standards, 5.9% are severely polluted. In Zhoushan and Lishui, the share of the average percentage of days on which air quality failed to meet safety standards ranges from 80% to 100%, while the average percentage of 23 cities is between 50% and 80%. Of the total number of days that failed to meet safety standards in the Yangtze River delta, more days have PM$_{2.5}$ as the major pollutant, taking a share of 80.0%. Days with O$_3$ and PM$_{10}$ as the major pollutant accounts for 13.9% and 5.8% respectively.

In 2013, the average percentage of days on which air quality in 9 cities at or above the prefecture-level in the Pearl River delta met safety standards ranges from 67.7% to 94.0%, and the average is 76.3%. Of the total number of days that failed to meet safety standards, 0.3% are severely polluted. In Shenzhen, Zhuhai and Huizhou, the share of the average percentage of days on which air quality failed to meet safety standards is higher than 80%, while the range for the other cities is between 50% and 80%. Of the total number of days that failed to meet safety standards in the Yangtze River delta, more days have PM$_{2.5}$ as the major pollutant, taking a share of 63.2%. Days with O$_3$ and NO$_2$ as the major pollutant accounts for 31.9% and 4.8% respectively.

Dust-haze

Visibility-based observation results by the China Meteorological Administration show that there are averagely 35.9 haze days in 2013 nationwide, up 18.3 days compared with 2012, and is the record high year since 1961. Fog and haze weathers occur frequently in Central and East China. In the central and southern part of North China as well as the northern part of region south of the Yangtze River, there are 50–100 days with fog or haze, and in some areas the number of such days exceeds 100.

Air quality-based monitoring results by the Ministry of Environmental Protection show that, in January and December of the year 2013, wide-range regional dust-haze pollution occurs twice in central and eastern China. Both pollution processes have the characteristics of large scope, long duration, high pollution level, fast rise of pollutant concentration and PM2.5 being the major pollutant.

In January, the dust-haze pollution process lasted for 17 days, resulting in 677 days of heavily or severely polluted weather in 74 cities, 477 days heavily polluted and 200 severely polluted. The Beijing-Tianjin-Hebei region and the neighboring areas are worse polluted, in particular the southern part of Hebei Province, and Shijiazhuang and Xingtai are the two worst hit cities.

During December 1–9, heavy dust-haze pollution process occurs in central and eastern China, resulting in 271 days of heavily or severely polluted weather in 74 cities, 160 days heavily polluted and 111 severely polluted. The Yangtze River delta, Beijing-Tianjin-Hebei region as well as the neighboring areas and part of northeastern China are worse polluted, and the Yangtze River delta is worst hit.

For the 256 cities at or above prefecture-level, the annual average values of SO_2, NO_2 and PM_{10} are assessed according to the Ambient Air Quality Standard (GB 3095–1996) in 2013, and 69.5% of the cities meet the standard. For annual average SO_2 concentration, 91.8% of the cities meet the standards, while 1.2% have sub-standard air quality worse than Level III. For NO_2, 86.3% of the cities meet the Level-I standard. And for PM_{10}, 71.1% of the cities meet the standards, while 7.0% worse than Level III.

Acid Rain

Acid rain frequency: In 2013, acid rain occurs in 44.4% of the 473 monitored cities, 27.5% have an acid rain frequency higher than 25%, and 9.1% higher than 75%.

Rainfall acidity: In 2013, cities with an annual average rainfall acidity lower than 5.6 (acid rain), 5.0 (relatively strong acid rain) and 4.5 (strong acid rain) account for 29.6%, 15.4% and 2.5% respectively. Compared with the previous year, percentage of cities with acid rain, relatively strong acid rain and strong acid rain has decreased by 1.1, 3.3 and 2.9 percentage points respectively.

In 2013, acid rain occurs mostly in the region along the Yangtze River, south of the middle and lower reaches of the river, including the major part of Jiangxi, Fujian and Hunan provinces, Chongqing Municipality, the Yangtze River delta, Pearl River delta, and southeast Sichuan Province. The area of acid rain accounts for 10.6% of China's total territory.

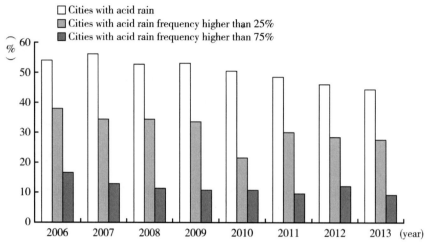
FIGURE II.4 Acid rain frequency nationwide during 2006–2013

Environmental Quality for Noise

Area-wide Environmental Quality for Noise

Cities at or above prefecture-level: Among the 316 cities monitored in daytime, area-wide environmental quality for noise in 76.9% of them is at Level I and Level II, 22.8% at Level III, 0% AT Level IV, and 0.3% at Level V. Compared with the previous year, the figure for Level II is down 1.8 percentage points, while cities at Level III is up 2.5 percentage points. No significant changes can be found with the other levels.

Among the 293 cities monitored in nighttime, area-wide environmental quality for noise in 48.5% of them is at Level I and Level II, 51.5% at Level III and Level IV, and 0% at Level V.

Key cities of environmental protection: In 2013, 113 cities are monitored in daytime and the equivalent sound level range of area-wide environmental quality for noise is 47.7–58.7 dB (A). Cities with area-wide environmental quality for noise at Level I and Level II account for 74.4%, Level III 25.6%, and 0% for both Level IV and Level V.

At the same time, 110 cities are monitored in nighttime and the equivalent sound level range of area-wide environmental quality for noise is 39.2–50.4 dB (A). Cities with area-wide environmental quality for noise at Level I and Level II account for 36.4%, Level III and Level IV 63.6%, and 0% for Level V.

Environmental Quality for Noise in Urban Functional Zones

In total, 17,696 points/times of monitoring are conducted in 2013 in all types of functional zones in cities at or above prefectural-level nationwide, half in the daytime and the other half at night. Of them, 91.1% monitored in the daytime are up to the standard, similar to the previous year, while the figure for the nighttime monitoring is 71.7%, up 2.1 percentage points from the previous year.

In key cities of environmental protection, 8,668 points/times of monitoring are conducted in all types of functional zones, half in the daytime and the other half at night. Of them, 90.7% monitored in the daytime are up to the standard while the figure for the nighttime monitoring is 67.9%.

TABLE II.9 Monitoring points reaching the standards in functional zones of cities at or above prefecture-level in 2013

Functional zones	Type 0		Type 1		Type 2		Type 3		Type 4	
	Day	Night	Day	Night	Day	Night	Day	Night	Day	Night
Up-to-standard points/times	68	48	1,838	1,502	2,556	2,278	1,677	1,517	1,923	997
Monitoring points/times	103	103	2,112	2,112	2,816	2,816	1,724	1,724	2,093	2,093
Up-to-standard rate (%)	66	46.6	87	71.1	90.8	80.9	97.3	88	91.9	47.6

TABLE II.10 Monitoring points reaching the standards in functional zones of key cities of environmental protection in 2013

Functional zones	Type 0		Type 1		Type 2		Type 3		Type 4	
	Day	Night	Day	Night	Day	Night	Day	Night	Day	Night
Up-to-standard points/times	36	26	792	626	1,337	1,148	859	757	907	387
Monitoring points/times	64	64	899	899	1,463	1,463	879	879	1,029	1,029
Up-to-standard rate (%)	56.3	40.6	88.1	69.6	91.4	78.5	97.7	86.1	88.1	37.6

Theme IV: Lands and Rural Environment

Conditions
The problem of farmland quality is outstanding, regional degeneration problem is severe, and the situation of rural environment is still grave.

Land Resources and Farmlands
According to the second national land survey, by the end of 2012, there are in total 646.4656 million ha of agricultural lands: farmlands 135.1585 million ha, forest lands 253.3969 million ha, and grasslands 219.5653 million ha. There are 36.907 million ha of construction lands, among which 30.1992 million ha are for urban, industrial and mining purposes.

In 2012, total farming area decreases by 402,000 ha because of constructions, disasters and ecological defarming, and increases by 321.8 ha because of land management and adjustment of agricultural structures, a net loss of 80,200 ha of farming lands.

Soil and Water Loss
According to the first national water resources survey and national survey on soil and water conservation, China now has a total eroded area of 2.9491 million square kilometers, or 30.72% of the national territory. Among them, 1.2932 million square kilometers are water erosion and 1.6559 million square kilometers wind erosion.

Theme V: Natural Ecological Environment

Conditions
The ecological environmental quality nationwide is in general stable.

Ecological Environmental Quality
In 2012, the ecological environmental quality nationwide is "moderate." Among the 2,461 counties, the number of counties with "high," "good," "moderate," "poor" and "bad" quality is 346, 1,155, 846, 112 and 2 respectively. Ecological environmental quality is mainly "good" and "moderate," accounting for 67.2% of the national territory.

Counties with "high" and "good" ecological environmental quality are mainly distributed in the regions south of the Qinling Mountains and the Huaihe River, the Greater Khingan Range, the Lesser Khingan Range, and the Changbai mountainous area in Northeast China. The "moderate" counties are mostly in the North China Plain, the western part of the Northeast China Plain, the central part of the Inner Mongolia and the Tibetan Plateau. The "poor" and "bad" counties are mainly in the northwestern region.

Biodiversity

In terms of biodiversity, China has all the types of terrestrial ecosystem, including 212 types of wood forest, 36 types of bamboo forest, 113 types of bushwood, 77 types of meadow, and 52 types of desert. China has a complex freshwater ecosystem, with 5 types of wetland (offshore and coastal wetland, river wetland, lake wetland, marsh wetland and artificial wetland), and 4 major marine ecosystems (Yellow Sea, North China Sea, South China Sea, and the Kuroshio Current sea area). In the offshore sea areas, there are typical marine ecosystems of coastal wetlands, mangrove forests, coral reefs, estuaries, gulfs, lagoons, islands, upwellings, sea grass beds, etc., and natural landscapes and relics of undersea ancient forest, marine erosion, marine accumulation geomorphy, etc. Artificial ecosystems mainly include farmland ecosystem, artificial forest ecosystem, artificial wetland ecosystem, artificial grassland ecosystem and urban ecosystem.

In terms of species diversity, China has 34,792 types of advanced plants, including 2,572 types of bryophytes, 2,273 pteridophytes, 244 gymnosperms, and 29,703 angiosperms. In addition, China has almost all the woody plants of the Temperate Zone. In China there are 7,516 types of vertebrates, including 562 types of mammals, 1,269 birds, 403 reptiles, 346 amphibians, and 4936 fishes. There are 420 types of rare and endangered animals in the national list of protected animal species, including giant panda, crested ibis, golden monkey, South China tiger and Chinese alligator. In China, more than 10,000 types of fungi have been identified.

In terms of the diversity of genetic resources, China has 538 classes and 1,339 types of cultivated crops, 1,000 economic tree species, 7,000 ornamental plants and 576 domesticated animals.

Natural Reserves

By the end of 2013, 2,697 natural reserves of different types at different levels have been established nationwide, covering a total area of 146.31 million ha. Among them, the terrestrial area is 141.75 million ha, accounting for 14.77% of China's total territory. There are 407 national natural reserves, covering an area of about 94.04 million ha.

Wetland

In 2013, 59 projects have been implemented for wetland protection, and the central government has appropriated funds for another 122 projects of wetland protection. There are 5 more Wetlands of International Importance, adding to a total number of 46 such wetlands in China. 131 national wetland parks (pilot) have been approved, and 300,000 ha of wetland area has been newly added.

TABLE II.11 The amounts and areas of natural reserves in 2013

Type	Amount		Area	
	Total amount	%	Total area (million ha)	%
Natural ecosystem	1,906	70.67	103.8507	70.98
Forest ecosystem	1,410	52.28	30.1332	20.60
Grassland and meadow ecosystem	45	1.67	2.1652	1.48
Desert ecosystem	37	1.37	40.8723	27.94
Inland wetland and aquatic ecosystem	344	12.75	29.9125	20.44
Marine and coastal ecosystem	70	2.60	0.7675	0.52
Wildlife	672	24.92	40.8465	27.92
Wild animal	525	19.47	38.9144	26.60
Wild plant	147	5.45	1.9322	1.32
Natural relics	119	4.41	1.6126	1.10
Geological relics	85	3.15	1.0610	0.73
Paleontological relics	34	1.26	0.5516	0.38
Total	2,697	100	146.3098	100

Invasive Alien Species

Currently there are about 500 invasive alien species in China. In the past decade, more than 20 of the world's worst invasive alien species have set their feet in China, and over 100 species have been causing large-area damages. The spartina alterniflora has spread an area of 35,995.2 ha along China's mainland coastline.

Forest Environment
Condition

Forest nationwide has been steadily developing with increase in both quantity and quality.

Forest resources

According to the 8th national survey on forest resources (2009–2013), national forest covers an area of 208 million ha, with the forest coverage rate being 21.63%, the total living wood growing stock 16.433 billion cubic meters, and forest growing stock 15.137 billion cubic meters. China is the world's No. 5 in forest area, No. 6 in forest growing stock, and No. 1 in artificial forest. Compared with the 7th national survey on forest resources (2004–2008), the forest area increases by 12.23 million ha, the forest coverage rate up 1.27 percentage points, and the net increase for living wood growing stock and forest growing stock is 1.52 billion and 1.416 billion cubic meters respectively. With increased forest area and improvement of structure and quality, the ecological functions of forest have been enhanced. Total biomass of forest vegetation nationwide reaches 17.002 billion tons, total carbon reserve 8.427 billion tons, annual conserved water 580.709 billion cubic meters, annual fixed soil 8.191 billion tons, annual conserved fertilizer 430 million tons, annual absorbed pollutant 38 million tons, and annual retained dust 5.845 billion tons.

Forest Biohazard

In 2013, an area of 7.205 million ha of forest nationwide is protected against major pests, less than 5‰ of the forest is affected by pests, and pollution-free pest control operation is conducted to 85% of the forest. Key pest hazards, including the pine wilt disease and fall webworm, is under control.

Grassland Environment
Condition
Grassland Resources

In 2013, grassland in China covers an area of about 400 million ha, accounting for 41.7% of the national territory. The area of grassland in the 12 western provinces or regions is 331 million ha, accounting for 84.2% of the total national grassland. The top 6 provinces or regions, namely Inner Mongolia, Xinjiang, Tibet, Qinghai, Gansu and Sichuan, have a share of 293 million ha, or 75.0% of the total. In South China, grassland is mainly grassy hills or slopes scattered in the hilly areas, and the total area is about 67 million ha.

Grassland Productivity

In 2013, natural grasslands nationwide have a total fresh grass yield of 1.0558121 billion tons, up 0.59% from the previous year and equivalent to 325.4292 million tons of hay. The grazing capacity is about 255.792 million sheep units, up 0.48% from the previous year. The 23 key provinces, provincial-level municipalities or regions nationwide yield 983.3337 million tons of fresh grass, 93.14% of the national total, up 0.41% from the previous year, equivalent to 307.817 million tons of hay. Their grazing capacity is about 242.0409 million sheep units, up 0.45% from the previous year.

Grassland Disaster

The year 2013 has totally 90 events of grassland fire disasters nationwide, among which 76 are small fires (Level I out of the four levels, the lowest), 13 big fires (Level II) and 1 serious fire (Level III). An area of 350.773 million ha is affected, causing an economic loss of 7.59 million *yuan*, 1 injury and no cattle loss. Compared with the previous year, there are 20 less incidents of grassland fires nationwide, of which 4 are big fires, 2 are serious fires and another 2 are disastrous fires (Level IV, the highest). The affected area decreases by 72.4%. Nationwide 36.955 million ha of grasslands are damaged by rats, accounting for 9.2% of total national grassland area and the share remains unchanged compared with the previous year. Nationwide 14.306 million ha of grasslands are affected by pests, accounting for 3.8% of total national grassland area and down 12.0% from the previous year.

Theme VI: Energy Condition

Condition

In 2013, national energy situation is in general stable, with steady supply and demand.

Production

In 2013, energy production is equivalent to 2.4 billion tons of standard coals, up 2.4% from the previous year. The production of raw coal is 3.68 billion tons (up 0.8% from the previous year), crude oil 210 million tons (up 1.8%), natural gas 117.05 billion cubic meters (up 9.4%), and electricity generation 5.39 trillion kilowatt hours (up 7.5%). During the year, coals imported reach 327 million tons, up 13.4%, crude oil 282 million tons (up 4.0%), and oil products 39.59 million tons (down 0.6%).

TABLE II.12 Production and growth rate of primary energy in 2013

Product	Unit	Production	Growth (%)
Total primary energy production	Million tons of standard coal equivalent	2,400	2.4
Raw coal	Million tons	3,680	0.8
Crude oil	Million tons	210	1.8
Natural gas	Billion cubic meters	117.05	9.4
Electricity generation	Billion kilowatt hours	5,397.59	7.5
Thermal power	Billion kilowatt hours	4,235.87	7.0
Hydropower	Billion kilowatt hours	911.64	5.6
Nuclear power	Billion kilowatt hours	110.63	13.6

Consumption

According to preliminary calculation, in 2013 total energy consumption in China is 3.75 billion tons of standard coal equivalent, up 3.7% from the previous year. Consumption of coal is up 3.7%, crude oil up 3.4%, natural gas up 13.0%, and electricity up 7.5%. Energy consumption for 10,000 yuan of gross domestic product nationwide is down 3.7%.

APPENDIX III

Air Quality Ranking for 2014 of Provincial Capitals and Central Government-Controlled Municipalities

2014 ranking	City	Annual average	2013 ranking
1	Haikou	2.44	1
2	Lhasa	3.23	3
3	Fuzhou	3.94	2
4	Kunming	4.18	4
5	Guiyang	4.32	5
6	Nanning	4.64	6
7	Shanghai	4.89	8
8	Nanchang	5.00	12
9	Guangzhou	5.08	7
10	Chongqing	5.56	9
11	Changsha	5.75	15
12	Hefei	5.95	13
13	Hangzhou	5.95	14
14	Xining	6.03	17
15	Hohhot	6.09	18
16	Yinchuan	6.19	11
17	Lanzhou	6.22	10
18	Urumchi	6.25	23
19	Changchun	6.41	16
20	Harbin	6.59	19
21	Chengdu	6.81	22
22	Wuhan	6.84	21
23	Nanjing	6.88	20
24	Taiyuan	7.15	25
25	Xi'an	7.27	29
26	Beijing	7.55	26
27	Shenyang	7.68	24
28	Tianjin	7.68	27

(cont.)

2014 ranking	City	Annual average	2013 ranking
29	Zhengzhou	7.79	28
30	Ji'nan	8.98	30
31	Shijiazhuang	10.54	31

Note: except Taiwan of China.
SOURCE: CHINA NATIONAL ENVIRONMENTAL MONITORING CENTRE, AIR QUALITY REPORT ON THE BEIJING-TIANJIN-HEBEI REGION, THE YANGTZE RIVER DELTA REGION, MUNICIPALITIES DIRECTLY UNDER THE CENTRAL GOVERNMENT, PROVINCIAL CAPITALS AND THE CITIES SPECIFICALLY DESIGNATED IN THE STATE PLAN. HTTP://WWW.CNEMC.CN/PUBLISH/TOTALWEBSITE/0666/NEWLIST_1.HTML.

As from January 1, 2013, 74 cities, including municipalities directly under the Central Government, provincial capitals, the cities specifically designated in the state plan and the cities above the prefecture level in the Beijing-Tianjin-Hebei Region, the Yangtze River Delta Region and the Pearl River Delta Region, were governed by the new *Ambient Air Quality Standard* (GB 3095–2012) and released the Air Quality Index (AQI) according to the *Technical Regulations on Ambient Air Quality Index (On Trial)* (HJ 633–2012).

The Air Quality Index (AQI) means an index for quantitative description of the condition of air quality; the larger its value is, the more severe the air pollution is and the greater the harm to human health is. AQI is classified into six grades: Grade I, excellent, Grade II, good, Grade III, mild pollution, Grade IV, moderate pollution, Grade V, heavy pollution, Grade VI, severe pollution. When the daily average concentration of PM2.5 reaches 150, 250 and 500μg/m^3, the AQI reaches 200, 300 and 500, respectively.

The data in this table's ranking are comprehensive of the ambient air quality index; it is a dimensionless index for describing the comprehensive condition of the quality of urban air; it gives overall considerations to the pollution levels of six pollutants, including fine particulate matter (PM2.5), inhalable particles (PM10), sulfur dioxide (SO_2), nitrogen dioxide (NO_2), ozone (O_3) and carbon monoxide (CO). The larger the comprehensive ambient air quality index is, the higher the comprehensive pollution level is; the single index of each pollutant is first calculated, and then the single indexes of six pollutants are added together to obtain the comprehensive ambient air quality index.

APPENDIX IV

List of Environmental Protection Laws and Regulations Released in 2014

TABLE IV.1 Laws and regulations

Names of law and regulation	Organization	Date of release
Regulations on the Management of Water Supply and Utilization in the South-to-North Water Diversion Project	The State Council of the People's Republic of China	Released on February 16, 2014, effective as from the date of promulgation
Law of the People's Republic of China on Environmental Protection (Revised in 2014)	The Standing Committee of the National People's Congress	Released on April 24, 2014, effective as from January 1, 2015
Decision of the State Council on Revising Some Administrative Regulations	The State Council of the People's Republic of China	Released on July 29, 2014, effective as from the date of promulgation

TABLE IV.2 Departmental rules

Rule names	Organization	Date of release
Circular of the General Office of the State Council on Strengthening Environmental Supervision and Law Enforcement	The General Office of the State Council	Released on November 12, 2014, effective as from the date of promulgation
Guiding Opinions of the General Office of the State Council on Improving the Rural Living Environment	The General Office of the State Council	Released on May 16, 2014, effective as from the date of promulgation

LIST OF ENVIRONMENTAL PROTECTION LAWS AND REGULATIONS 265

TABLE IV.3 Regulatory documents

Document names	Organization	Date of release
Circular of the General Office of the Ministry of Environmental Protection and the General Office of the Ministry of Housing and Urban-Rural Development on the Printing and Distribution of the Regulations on the Management of Archives Concerning Major Special Science and Technology Projects for Water Body Pollution Control and Abatement (Trial)	Ministry of Environmental Protection/Ministry of Housing and Urban-Rural Development	Released on January 10, 2014, effective as from the date of promulgation
Circular of the General Office of the Ministry of Environmental Protection on Implementing the Action Plan for Air Pollution Prevention and Control and Exercising Strict Control over Access to Environmental Impact Assessment	Ministry of Environmental Protection	Released on March 25, 2014, effective as from the date of promulgation
Circular of the General Office of the Ministry of Environmental Protection on Releasing the Catalogue of Hazardous Chemicals as the Environmental Management Priorities	Ministry of Environmental Protection	Released on April 3, 2014, effective as from the date of promulgation
Circular of the General Office of the Ministry of Environmental Protection on Strengthening Prevention and Control of Cyanobacterial Bloom in Key Lakes	Ministry of Environmental Protection	Released on June 27, 2014, effective as from the date of promulgation

TABLE IV.3 Regulatory documents (*cont.*)

Document names	Organization	Date of release
Circular of the National Development and Reform Commission, the Ministry of Finance, the Ministry of Environmental Protection on Relevant Issues Including the Adjustment of the Standard for Pollutant Charge Collection	Ministry of Finance/ Ministry of Environmental Protection/National Development and Reform Commission (including the former State Development Planning Commission, the former State Planning Commission)	Released on September 1, 2014, effective as from the date of promulgation
Circular of the National Development and Reform Commission, the Ministry of Environmental Protection and the National Energy Administration on the Printing and Distribution of the Action Plan for Coal Power Energy Conservation, Emission Reduction Upgrading and Transformation (2014–2020)	Ministry of Environmental Protection/National Development and Reform Commission (including the former State Development Planning Commission, former State Planning Commission)/National Energy Administration	Released on September 12, 2014, effective as from the date of promulgation
Circular of Six Ministries including the Ministry of Environmental Protection, the National Development and Reform Commission and the Ministry of Public Security on the Printing and Distribution of the 2014 Implementation Plan for the Phase-out of Yellow-label Cars and Old Cars	Ministry of Public Security/ Ministry of Environmental Protection/the National Development and Reform Commission (including the former State Development Planning Commission, the former State Planning Commission)	Released on September 15, 2014, effective as from the date of promulgation

TABLE IV.3 Regulatory documents (*cont.*)

Document names	Organization	Date of release
Circular of the General Office of the Ministry of Environmental Protection and the General Office of the Ministry of Public Security on Production, Use and Environmental Supervision and Innocent Destruction of Precursor Chemicals	Ministry of Public Security/ Ministry of Environmental Protection	Released on October 17, 2014, effective as from the date of promulgation
Circular of the Ministry of Environmental Protection, the Ministry of Education, the Ministry of Science and Technology etc. on Strengthening the Management of the Utilization of, Access to and Benefit Sharing of Biological Genetic Resources in External Cooperation and Exchange	Ministry of Education/ Ministry of Science and Technology/Ministry of Environmental Protection	Released on October 28, 2014, effective as from the date of promulgation
Circular of the General Office of the Ministry of Environmental Protection on the Printing and Distribution of a Series of Technical Guidelines for the Ecological Environmental Protection of Rivers and Lakes	Ministry of Environmental Protection	Released on December 23, 2014, effective as from the date of promulgation
Circular of the General Office of the Ministry of Environmental Protection on the Printing and Distribution of the List of Enterprises Subject to National Focused Supervision 2015	Ministry of Environmental Protection	Released on December 31, 2014, effective as from the date of promulgation

APPENDIX V

A Letter to the Government about Mandatory Protective Book Jackets

The Friends of Nature is a non-governmental environmental protection organization. We pay attention to the issue of solid urban wastes and hope that urban waste reduction, resource saving and recycling can be promoted. According to a survey conducted by the Friends of Nature, at present, it is common for teachers to mandatorily require students to add protective book jackets in Beijing's primary schools, which not only increases an extra burden of cost on parents, but it also causes severe wasting of resources. Therefore, the Friends of Nature calls on primary schools in Beijing to cancel the regulations and requirements for mandatorily demanding addition of protective book jackets.

In early 2014, the Friends of Nature carried out an online questionnaire survey regarding the mandatory requirement of adding protective book jackets, in which the parents of students at primary schools in Beijing were respondents. In this survey, 97 feedbacks were received from parents, involving 62 primary schools in 8 urban districts, among which about 90% of the primary schools mandatorily required students to add protective book jackets. Among these 62 primary schools, about 30% required students to adopt finished plastic protective book jackets, 10% required students to use white paper as protective book jackets, 35% required students to internally and externally adopt white paper and ready-to-use plastic protective book jackets, respectively—every book was covered by two layers of jackets—15% did not impose restrictions on the materials used for book jackets, while only 7 primary schools, including 2 private primary schools in Changping District, did not require any addition of protective book jackets, which accounted for only 10% in this survey.

In the schools which required the purchase of ready-to-use protective book jackets, in order to keep their protective book jackets neat, nice-looking, uniform and hygienic, about 50% of the teachers required students to replace them every semester in case they were stained or damaged.

Therefore, most of the parents needed to purchase a large quantity of protective book jackets for their children, and spending 100,000–300,000 yuan every semester. Parents generally objected to such mandatory provision and thought that this increased the economic burden on them, wasted enormous resources and was highly environmentally unfriendly, in particular, the provision that double-layer protective book jackets were required was extremely unreasonable; parents hoped that school teachers would allow parents and students to make free choices and respect the autonomous right of the students.

According to our estimation, if schools mandatorily require that ready-to-use protective book jackets be used and replaced every semester, and it is assumed that every student needs at least 15 protective book jackets—6 books and 9 exercise books—every semester, and double-layer book wrappers are not considered, one student will consume 180 protective book jackets during his/her six-year schooling period at primary school. Currently, there are about 800,000 primary school students in Beijing; according to the above estimation, about 24 million protective book jackets are used by primary school students every year in Beijing. The quantity of resources thus consumed is alarming! Moreover, most of the used-up book wrappers are made of plastic, so if they are abandoned after use every semester, they will become difficult-to-degrade wastes and so the environment will be polluted as a result.

Therefore, the Friends of Nature has publicly made a special appeal to Beijing's education department, the persons in charge of and class teachers at primary schools as the new fall semester is drawing near:

I. The Friends of Nature hopes that educators can change their thinking and shift their focus from the pursuit of neat, uniform and nice-looking protective book jackets to frugality, environmental friendliness and practicality; from specifying the mandatory provision to respecting the right of students and parents to make free choices, and surrender to students the decision-making power concerning whether to use protective book jackets or not and the selection of materials of protective book jacket, and make it so that educators no longer set unified hard-and-fast rules.

II. It should be stressed that plastic protective book jackets are reusable. The Friends of Nature hopes that school teachers no longer require mandatory replacement of protective book jackets every semester, and that they should educate students to, as far as possible, reuse the used protective book jackets and extend their service life by taking good care of their books and cleaning them every day.

III. The Friends of Nature hopes that school teachers can deal with the "minor matter"—protective book jackets—in a more flexible way by cultivating and educating students with respect to their way of cherishing books, their environmental awareness and their ability to be practical. In the questionnaire survey, more than 60% of the parents said that using their old possessions again, avoiding the wasting of material, saving things and taking care of their possessions, being mindful of environmental protection and what is beneficial to the education of children are their priorities when selecting protective book jacket. Conventional waste and old paper is the preferred material for book wrappers for them.

Primary school is the important age for children to develop their outlook on life and living habits. the idea of treasuring properties, and saving and protecting the environment should be part of primary school education. We believe that education which goes deep into the hearts of people should be the result of an invisible and formative influence from daily life rather than the rigid instillation of slogans and concepts. We hope that Beijing's educators can start from the "minor matter"—protective book jacket—and from changing their notions and behaviors, and act together with the children to achieve common growth, pay attention to environmental protection and jointly create a bright future.

    The Friends of Nature
    August 21, 2014

APPENDIX VI

Friends of Nature Suggestions for Amending the *Air Pollution Prevention and Control Law*

Dear Legislative Affairs Office of the State Council:

We are the Friends of Nature, an environmental protection organization in Beijing. We have always been paying great attention to solving the problems regarding air pollution. Recently, we have learned that public opinion about the revision of the *Air Pollution Prevention and Control Law* is being solicited. We are very happy about this revision and hope that our practical actions and experience can contribute constructively to this revision.

Reading the exposure draft carefully, we believe that, in general, many improvements should be made to it. Air pollution is a complicated issue. In making the *Air Pollution Prevention and Control Law*, overall considerations should be given to various complicated scenarios. We suggest that opinions should be extensively solicited from experts in relevant fields and from the general public; a full feasibility study should be conducted; appropriate measures for addressing relevant issues should be taken; government responsibilities should be specified; market means should be adopted; importance should be attached to social supervision so as to make a law which can be effectively enforced and that can really solve the problems in air pollution prevention and control.

Specifically, first, the exposure draft has no adequate provisions relating to information disclosure and public participation. The *Environmental Protection Law* deliberated and adopted in April this year dedicates a special chapter to dealing with "information disclosure and public participation." According to the first provision of this chapter, citizens, legal persons and other organizations enjoy, according to laws, the right to obtain environmental information, participate in and supervise environmental protection. As a single legislative act in the area of air pollution, the *Air Pollution Prevention and Control Law* should be consistent with the *Environmental Protection Law* and should provide a special chapter containing detailed and comprehensive provisions for the disclosure of information and public participation in the area of air pollution. In reference to the *Environmental Protection Law*, it is suggested that this chapter should include the following contents:

1. Citizens, legal persons and other organizations enjoy, according to laws, the right to obtain environmental information relating to air pollution, participate in and supervise the protection of the air.

2. The competent departments for environmental protection regarding the air and other departments for the supervision and administration of environmental protection at the various levels of the people's governments should disclose environmental information according to laws, improve the procedure for public participation, and provide convenience to citizens, legal persons and other organizations for participating in and supervising environmental protection.
3. The major air pollutant discharging units should truthfully make public the names of their main pollutants, discharge methods, discharge concentration and the total quantity of their discharge, the amount of and kind of discharge beyond the standard level, the construction and operation of their pollution prevention and control facilities, to accept public supervision.

    In our opinion, disclosing discharge information to the general public on a real-time basis is an important means of social supervision for urging enterprises to legally discharge pollutants. Therefore, we suggest that the *Air Pollution Prevention and Control Law* should be consistent with the *Environmental Protection Law* and should provide for major pollutant discharging enterprises to be duty-bound to make their real-time discharge information public and to accept social supervision.
4. With regard to the construction projects for which an environmental impact assessment report is prepared according to laws, construction units should, at the time of preparing such a report, provide relevant information to the people who may be affected, fully solicit their opinions, respond to their opinions, and explain the reasons for not considering their opinions.
5. The department in charge of examining and approving the environmental impact assessment document concerning construction projects should, after receiving the *Environmental Impact Assessment Report on a Construction Project*, disclose its full text except the matters involving state secrets and business secrets. Where it is found that no public opinions on construction project have been completely solicited, construction units shall be ordered to solicit those opinions.
6. The general public, NGOs for environmental protection and other organizations are entitled to report any environmental pollution behavior to the competent department for environmental protection; where the competent department for environmental protection does not actively perform its duties within ten working days after receiving such a report, the whistle-blower is entitled to sue the competent department for environmental protection at the people's court, and require it to actively perform its duties; the whistle-blower can also directly sue the environmental polluter of the air at the people's court.
7. For the behavior of polluting the air and harming social public interests, relevant NGOs can initiate legal proceedings according to the laws.

First, air pollution is an environmental issue very closely connected to the people's lives. We hope that the new *Air Pollution Prevention and Control Law* will make more progress than the new *Environmental Protection Law* in information disclosure and public participation has so far made, and that it will set forth comprehensive and detailed stipulations for the disclosure of information and for public participation in the field of air pollution prevention and control. Information disclosure and public participation should be carried out in order to improve social supervision over the behavior of polluting the air so as to contain the increasing severity of air pollution and gradually realize our "blue sky dream."

Second, with respect to standard-setting, in order to protect the air, the local authorities should be encouraged to develop discharge standards that are stricter than the national standards; it is suggested that local standards should not be submitted to the State Council for approval and should only be filed with the State Council if local standards are stricter than the national standards.

Therefore, it is suggested that Paragraph 3 of Article 9 should be deleted. Paragraph 3 of Article 9 provides that where local discharge standards for air pollutants from motor vehicles and ships as developed by the people's governments of provinces, autonomous regions and municipalities directly under the Central Government are stricter than the national discharge standards, these local discharge standards should be submitted to the State Council for approval.

Third, in general, the provisions of the exposure draft concerning legal liability are less comprehensive and not as strict, so we worry that it cannot change the current situation in which there is no cost or low cost for violating the laws. It is suggested that the legal liabilities of liability subjects should be improved in an all-round way, and the legal liabilities of relevant subjects should be specified if there are legal obligations and duties of relevant subjects; furthermore, it is suggested that a stricter liability assumption system and a multistep punishment system should be established; for example, a higher amount of fines should be set or stricter punishment measures, such as closing down enterprises, should be adopted for repeated violations of the laws.

We have also summarized the public opinions from the advocacy activity of nongovernmental participation in legislation under the theme "I present suggestions for the *Air Pollution Prevention and Control Law*" as initiated by the Friends of Nature in April this year. We have also made a comparison between the *Environmental Protection Law* and the *Air Pollution Prevention and Control Law* (exposure draft for draft amendment) for reference.

Given time constraints, with careful study and discussion, we put forward the above superficial suggestions for reference, while the opinions on revising specific provisions need to be studied and discussed more carefully given a sufficient amount of time. We hope that there will be more opportunities for public participation in revising the *Air Pollution Prevention and Control Law* in the future, we will also pay close and constant attention to the revision of this law. We hope that we can offer scientific suggestions for legislation on the basis of our actions. Finally, we hope that, based on legislative work led by the legislative body, with participation by the people from various sectors of the society, a law capable of effectively solving the problem of air pollution which draws public attention can be made!

>The Friends of Nature
>September 30, 2014

Note: We are improving the provincial platform for disclosing the self-monitoring information on the pollution sources under intensive control by the State; the Ministry of Environmental Protection has also issued an official document to require the disclosure of real-time monitoring information on pollutants from the pollution sources under intensive control by the State through the unified provincial platform as of January 1, 2014.

APPENDIX VII

Friends of Nature Continues to Urge Relaxation of Eligibility Requirements for Litigants in Environmental Public Interest Suits

Dear members of the Standing Committee of the National People's Congress, the Law Committee of the National People's Congress, the Legislative Affairs Committee of the Standing Committee of the National People's Congress:

Today is World Earth Day and World Law Day. The Friends of Nature has been informed that the *Environmental Protection Law* is being deliberated in the Standing Committee of the National People's Congress for the fourth time. The draft for the fourth deliberation has a provision relating to the subject qualification for environmental public interest litigations. According to the provision, where pollution of the environment is under way, or the ecology is being destroyed and social public interests are being harmed, the non-governmental organizations which satisfy the following conditions can file litigations at the people's court:

> (I) They are registered with the civil affairs department of the people's government above the level of the city divided into districts according to the laws; (II) they have specialized in public interest activities relating to environmental protection for more than five consecutive years and enjoy a good reputation.

The provision presents a significant improvement compared with last year's drafts of the environmental protection law for the second and third deliberations, and it extends the subject qualification for environmental public interest litigation to the non-governmental organizations registered with the civil affairs department above the level of the city divided into districts. This indicates that the revision of the *Environmental Protection Law* carries out the new spirit of the Third Plenary Session of the 18th Central Committee of the Communist Party of China to some extent, and reflects China's determination regarding the fight against pollution. This shows that China has decided to solve its environmental problems by orderly promoting the institutionalization of judicial and social forces.

According to the *Decision of the Central Committee of the Communist Party of China on Some Major Issues Concerning Comprehensively Deepening the Reform*, in order to build up an ecological civilization, it is essential to develop a systematic and complete institutional system for that kind of civilization, adopt the strictest systems of protection of the source, damage compensation and accountability, improve the system of environmental governance and ecological remediation, and protect the ecological environment through institutions. Building up institutions is an important guarantee for pushing forward the construction of an ecological civilization and serves as the fundamental way to achieve the dream of a beautiful China. Improving the environmental public interest litigation system is precisely what is needed to build up a systematic and complete institutional system for an ecological civilization and it is an important institutional guarantee for effectively holding the environmental polluters and ecological destroyers accountable for their actions.

However, as a public interest organization for environmental protection dedicated to promoting the rule of law in the environmental field in China, we must say that the legislative proposal is still too strict with the subject qualification for environmental public interest litigation.

First, it limits the non-governmental organizations qualified for filing public interest litigations to those above the level of the city divided into districts, while there is no basis for such a limitation and the regional characteristics of China's environmental problems are not fully considered in a limitation of that kind. The environmental problems often show obvious regional characteristics, and they first affect the environmental interests of local people; therefore, the grass-roots environmental protection organizations should be fully granted the right to file environmental public interest litigations.

Second, the actual situation of the non-governmental organizations for environmental protection in China is not taken into full consideration. Given the difficulty in registration, many environmental protection organizations are currently engaged in environmental protection in the capacity of environmental protection volunteers, and a considerable number of registered environmental protection organizations are registered with the civil affairs department at the district level, and even some environmental protection organizations have to be registered with the department for industry and commerce.

Third, "above the level of the city divided into districts" does not take into account the special circumstance of the municipality directly under the Central Government, and arouses operational controversies and difficulties.

Finally, what are the criteria for judging "a good reputation"? Who makes that judgment? It is very easy for such highly flexible regulations to cause an abuse of the discretionary power.

The Friends of Nature once again calls for permitting all legally-registered non-governmental organizations for environmental protection to become the subjects qualified for filing the environmental public interest litigations, and fully encouraging various social forces to orderly fight against pollution in an institutionalized way!

Friends of Nature
April 22, 2014

APPENDIX VIII

A Petition Calling for Releasing for "Soliciting Public Opinions and Ensuring that the Newly Revised *Standards for Controlling Pollution from Residential Waste Incineration* by July 1, 2014"

Recent sustained haze across China is an environmental difficulty which troubles every citizen and the government, so solving the haze problem has become an urgent task. The persistent organic pollutants from household refuse incineration plants such as sulfur dioxide, oxynitride and dioxin can exert a continuous adverse impact on the environment and on human health. The pollutant discharge from refuse incineration plants as an important cause of air pollution should be effectively controlled and supervised.

According to a scholar from the China Urban Construction Design & Research Institute, the estimated dioxin air discharge from 60–70 household refuse incineration plants across China in 2007 was 157.93g TEQ, representing a significant increase compared with the 125.8g TEQ in 2004. As of August 2013, about 150 household refuse incineration plants had been under operation throughout China, except Hong Kong, Macao and Taiwan; the number of refuse incineration plants in China is expected to exceed 300 by the end of the 12th Five-Year Plan period. Continuous construction of refuse incineration plants across China will certainly increase the total discharge and toxicity of pollutants.

Worryingly, the pollutant discharge from refuse incineration plants in China is still governed by the 2001 *Household Refuse Incineration Pollution Control Standard* (2001 Standard), which obviously lags behind the European standard and has a difference of 10 times with the European standard in terms of the limit for dioxin concentrations. This is obviously not able to ensure effective control of pollutant discharges from refuse incineration plants. The introduction of the new control standard was delayed after opinions on it were solicited in 2010! Three years later, the second exposure draft of the *Household Refuse Incineration Pollution Control Standard* (the second exposure draft) was issued on December 27, 2013.

Therefore, we once again call on the national department for environmental protection to speed up the introduction of the new *Household Refuse Incineration Pollution Control Standard* so that we will not wait for three more years!

Moreover, in response to the second exposure draft, we believe that a stricter standard and a more extensive exposure draft are required. First, the limits for many pollutants in the second exposure draft are looser than those in the exposure draft of 2010; for example, the limits for oxynitride in the exposure draft of 2010—a 1-h average of 250 mg/m³ and a 24-h average of 200 mg/m³—have been extended to a 1-h average of 300 mg/m³ and a 24-h average of 250 mg/m³; the determined average of mercury and its compound, and cadmium and thallium and their compounds has been extended from 0.05 mg/m³ to 0.1 mg/m³. Obviously, such a standard cannot effectively control the pollutant discharge from refuse incineration plants. Second, all of the units whose opinions are to be solicited as listed in the opinion solicitation letter are government departments, and no opinions are solicited from the general public, environmental protection organizations and the people from other sectors of the society, so there are great limitations.

In view of this, we bring forward the following initiatives and solemnly call on the national department for environmental protection:

I. With regard to the *Household Refuse Incineration Pollution Control Standard* (the second exposure draft), opinions should be solicited from the general public, environmental protection organizations and the people from other sectors of the society, the solicitation of opinions should not be limited to the list specified in Annex 1, and a standard modification hearing for the general public should be held.

II. Actions should be taken to improve the working efficiency and ensure that the new, stricter *Household Refuse Incineration Pollution Control Standard* can be introduced on July 1, 2014.

We believe that introducing a strict standard is the first step for controlling the pollutant discharge from refuse incineration plants. Only when the supervision over household refuse incineration plants is intensified and full information disclosure and public participation is achieved can clean operation of refuse incineration plants be promoted and the environmental pollution from refuse incineration and its impact on human health be mitigated at the source.

Environmental protection organizations will present the modification opinions on the *Household Refuse Incineration Pollution Control Standard* (the second exposure draft) prior to January 15, and the national department for environmental protection is requested to carefully read them and make a reply.

January 7, 2014

Organizations which bring forward initiatives:
Wuhu Ecology Center
The Friends of Nature
Nature University
Shanghai Rendu Marine Public Interest Development Center

Aifen Environmental Protection Science & Technology Consulting Service Center
Guizhou Gaoyuan Renewable Resource Recycling (Co., Ltd.)
Green Earth Environmental Protection Science & Technology (Co., Ltd.)
Eco Canton
Xiamen Greencross
Guizhou Greenhome Environmental Protection Volunteer Association
Zero Waste Alliance
Changsha Lvdong Community Environmental Protection Service Center
Zhengzhou Environmental Protection Association
Henan Green Central China Environmental Protection Association
Nanjing Green Stone Environmental Action Network
Kunshan Lucheng Environmental Protection Volunteer Service Association
Xi'an Green Origin Environmental Publicity & Educational Development Center
Friends of Nature Shanghai Ecological Protection Association
Chengdu Roots & Shoots Environmental Cultural Exchange Center
Envirofriends Science & Technology Research Center
Fujian Green Home Environment-Friendly Center
Chongqing HengAo Environmental Protection Industry Development Co., Ltd.
Green Educational Working Committee of the Tianjin Environmental Science Society (Friends Of Green China Tianjin)

Individuals:
Lu Xiaoqin
An Xincheng

January 7, 2014

APPENDIX IX

2014 Winners of Ford Motor Conservation & Environmental Grants, China

Shanghai, November 28, 2014—Today, the award ceremony for the 2014 Ford Motor Conservation & Environmental Grants, China under the theme "You Are an Environmental Protection Hero" and the Green Travel Forum were successfully concluded in Shanghai. Twenty-nine environmental protection organizations across China received 29 awards under three categories, which were Natural Environmental Protection—Pioneer Award, Natural Environmental Protection—Communication Award and Community Participation Creativity Award, and they were granted 2 million yuan in a money award. After the award ceremony, Mrs. Li Ying, Vice-president of Communication & Public Affairs, Ford Motor (China) Co., Ltd. and Dr. Zhang Li, wildlife protection expert and judge of Ford Motor Conservation & Environmental Grants, China, Mr. Zhang Ning, Founder and Council President of Shanghai Yiyou Youth Service Center, Mr. Zhu Dajian, Director of the Sustainable Development & Management Research Institute of Tongji University, Mr. Li Yi, Secretary General of the Mobile Internet Industry Union, attended the Green Travel Forum to jointly discuss the current situation of green travel and the prospect of green, environmentally-friendly and sustainable development.

The award-winning projects involving the 2014 Ford Motor Conservation & Environmental Grants, China:

1. Natural Environmental Protection—Pioneer Award
The First Award
Project name: Yangtze River Source Ecological Rescue Action
Project director: Yang Xin
Project location: Sichuan Green River Environmental Protection Promotion Association

The Second Award
(1) Project name: Green Zhejiang "Our Water, Our Actions"
Project director: Xin Hao
Project location: Zhejiang Green Scientific, Technological and Cultural Promotion Association
(2) Project name: Adopt the Advocacy Strategy to Protect Grassland Ecology
Project director: Ding Wenguang

Project location: Western Environmental & Social Development Center, Lanzhou University

The Third Award
(1) Project name: Environmental Law Earth Action Project
Project director: Zhou Xiang
Project location: Green Anhui Environmental Consulting Center
(2) Project name: Water Environmental Protection Invention Example
Project director: Wu Guohua
Project location: Shanghai Nanyang Model School Innovation Association
(3) Project name: Green Startups
Project director: Ge Yong
Project location: Institute for the Environment and Development

Nomination Award
(1) Project name: General Survey on Safeguarding The Fuhe River
Project director: Ke Zhiqiang
Project location: Wuhan Green City of Rivers Environmental Protection Service Center
(2) Project name: Community Co-management, Fishing Village Self-governance
Project director: Wang Meng
Project location: Sanya Blue Ribbon Ocean Protection Association
(3) Project name: Scientific Protection of the Yehe Wetland
Project director: Li Jianping
Project location: Xibaipo Bird Association, Pingshan County

2. Natural Environmental Protection—Communication Award
The First Award
Project name: "Home" Trilogy Micro Film for Public Interest
Project director: Qiao
Project location: "Shoot to Protect" camera crew

The Second Award
(1) Project name: Projectionist on the Road
Project director: Long Shan
Project location: HOME Community Volunteer Service Center, Changsha
(2) Project name: Environmental Legal Assistance Action Network
Project director: Wang Canfa, Tian Wenpeng
Project location: the Center for Legal Assistance to Pollution Victims, China University of Political Science and Law

The Third Award
(1) Project name: Environmental Protection Education Base of the Children's Museum
Project director: Zhang Ning
Project location: Shanghai Yiyou Youth Service Center
(2) Project name: Protect Saunders's Gull
Project director: Liu Detian
Project location: Saunders's Gull Conservation Society of Panjin City
(3) Project name: Coastal Central Ecological Survey
Project director: Hu Liujun
Project location: Lianyungang Newspaper Media Group

Nomination Award
(1) Project name: Green Water Action
Project director: Sheng Jian
Project location: Social Work Service Center, Nanjing Love Journey
(2) Project name: South-to-North Water Diversion Cherishing Water Communication Action
Project director: Yan Fusheng
Project location: Great Love Warm Sun Society, ChangAn District, Shijiazhuang City
(3) Project name: Environmental protection TV program Green Home
Project director: Jia Dai
Project location: Hulunbeier Municipal Environmental Protection Publicity & Education Center

3. Community Participation Creativity Award
(1) Project name: Mitigate Non-point Source Pollution from Cultivation in Paddy Field
Project director: Lu Chunming
Project location: Pingbian Natural Protection Association
(2) Project name: Youth's Joint Efforts, Low Carbon Campus Project
Project director: Wang Jian
Project location: (China) Youth Climate Action Network
(3) Project name: HOME 36 Houses Space Joint Laboratory
Project director: Long Shan
Project location: HOME Community Volunteer Service Center, Changsha
(4) Project name: Fertile Soil Young Returnee Support Project
Project director: Hao Guanhui
Project location: Nurture Land
(5) Project name: Where to Go—River ICU Project
Project director: Dai Guangliang

Project location: Guangzhou Association for the Promotion of New Life Environmental Protection

(6) Project name: Waste Reduction, Green Life

Project director: Wang Zhiqin

Project location: Shanghai Longnan Green Housewife Environmental Protection Science & Technology Studio

(7) Project name: Guoren Dutch Treatment Community Canteen

Project director: Liang Youfei

Project location: Chongqing Guoren Dutch Treatment Culinary Culture Co., Ltd.

(8) Project name: Dongzhaigang Mangrove Forest Bee-keeping Project

Project director: Feng ErHui

Project location: Hainan Dongzhaigang National Nature Reserve Authority

(9) Project name: Tongxin Huhui Community Donation Collection Point

Project director: Sun Heng

Project location: Beijing Fellow Worker's Home Cultural Development Center

(10) Project name: Procapra Przewalskii Ecological School

Project director: You Luqing, Zhao Haiqing

Project location: Haiyan Zhabula Ecological Protection Association

(11) Project name: Kidney of the Earth

Project director: Wu Hao

Project location: Enactus Team, Zhuhai Branch, Beijing Normal University

APPENDIX X

2014 Winners of Best Environmental Report Award in China

On the afternoon of May 27, 2014, the appraisal and selection result of the 2014 (5th) Best Environmental Report Award jointly sponsored by Chinadialogue, Netease New Media Center and Renmin University of China was announced. Besides a money award, some first award winners will, at the invitation of the German Ministry of Foreign Affairs and Chinadialogue, go to Germany for a one-week exchange visit.

The Second Award
Tears of Water Swans in Dongting Lake—20 Small Swans Died in Dongting Lake, Lifeng, Changsha Evening News
How PJM Mode is Operated in the USA, Liu Laya, Southern Energy Observer
Chasing the Haze, Yan Hao, Huang Fang, Shi Yi, Huang Zhiqiang, Oriental Morning Post
Anti-green Campaign in the Oil Industry, Lu Minghe, Yuan Duanduan, Feng Jie, Li Yifan, Gong Junnan, Southern Weekend
Air Improvement Policy Series, Wang ErDe, 21st Century Business Herald
"Binding" the Flying Ashes in Wuhan, Lu Zongshu, Southern Weekend
The "Underground" History of Underground Water Control, Feng Jie, Southern Weekend
After 7.5 Billion was Put into the Three River Source Region, Li Jing, Oriental Outlook
Experts Analyzed the "Prescriptions" for Underground Water Pollution Control in North China, Gan Xiao, Chinese Science News
The Underground Water Crisis in Haolebaoji, Xu Zhihui, China Newsweek

The Second Award for Citizen Journalist
Huang Yunguo (@shennonggengzhe)
Since a farmer in Shennongjia, Hubei Province used a microblog through his mobile phone, the environmental problems in Shennongjia have been continuously exposed: a number of problems including deforestation for planting tobacco, allowing air traffic without environmental impact assessment of airport development, a dilemma in protection of the Dajiuhu wetland, mass casualties of wild animals resulting from excessive animal traps on mountains, the construction of a hydropower station in the reserve area, constant poaching of golden pheasants, an animal under first-grade State protection, have all drawn wide attention. Meanwhile, he has also actively volunteered

to remove the animal traps on mountains as well as the rural garbage, and to explore the development of ecological tourism; moreover, he has become a local environmental supervisor.

Chen Haobo et al. (@dongtingshouhuzhe)
One of the strongest forces is a finless porpoise protection team composed of fishermen, and is the group most familiar with the finless porpoise endangered species in China. They have protected finless porpoises and the ecology of Dongting Lake since 2003. Their microblog is less used, but the Lake Tour Diary, which is occasionally released, is lively and touching thanks to the expression of a simple feeling about non-governmental environmental protection. They are demonstrating the real value of environmental protection to society in the most rustic way.

Single Special Award
Best Investigative Reporting Award: *Cadmium Disease is Looming Large*, Liu Hongqiao

The speech from the judging panel at the award ceremony of the *New Century Weekly* within Caixin magazine: This award is granted to both Liu Hongqiao, the author of "Cadmium Disease is Looming Large," and the Environmental and Health Reporting Team of the *New Century Weekly* within Caixin magazine. Since 2011, Caixin magazine has continuously published great reports focusing on cadmium pollution and revealing a chain of disasters from cadmium-induced soil pollution to cadmium-induced rice pollution and then cadmium disease. The journalists of the *New Century Weekly* within Caixin magazine have been given China's Best Environmental Report Award many times due to these reports.

"Cadmium Disease is Looming Large," written by Liu Hongqiao is a case which vigorously combines the traditional investigative report with the environmental science report. In this report, vivid but moderate words are used to uncover the pain caused to victims due to incidents of environmental cadmium pollution; a complete chain of cadmium damage to the environment and health is unprecedentedly shown to the general public through down-to-earth investigations and detailed arguments. During their interviews, the journalists went deep into "cadmium villages" and consulted a huge amount of Chinese and English scientific literature. This abstruse literature, seldom accessible to the general public, and in-depth interviews have enabled the journalists to provide the whole picture of "cadmium disease being looming large" and have ensured that this report is scientific.

Best Impact Award: *The Arsenic Pollution in Underground Water is Endangering 20 Million Chinese People*, Xuan Jinxue, China Youth News
The speech from the judging panel at the award ceremony: A paper published in the U.S. *Science* magazine stated that 19.58 million Chinese people lived in the high-risk

areas with underground water containing excessive arsenic in China. It is a giant figure, but the situation behind this figure is little known to the outside world. With the story of villager Wu Zhiqaing at the age of 47 as the victim, the journalists have uncovered the shocking and heartbreaking story behind this great figure.

The journalists have proceeded from the individuals whose destiny has been changed by arsenic pollution in underground water to reveal the cause of arsenic pollution in underground water, its development, distribution and the severity of the ensuing problems in China. Among relevant reports, this report is most detailed, in-depth and presents rich multidimensional content, so it is highly influential.

Best Emergency Report Award: *Pollution was Discharged into Underground Water*, Gao Shengke, Xu Jing, He Tao
The speech from the judging panel of Caixin magazine at the award ceremony: When there was worry about the underground water pollution in Shandong and the incident involving direct discharge of pollution into underground water by enterprises in Weifang went viral on the microblog, but many media failed to find any evidence of these problems, the journalists of Caixin magazine conducted investigations for nearly half a month and found proof of a scene of pollution discharge from key provincial enterprises into the underground water in Pingdingshan City, Henan Province, a city one thousand miles away.

According to on-the-spot investigations, underground water pollution in the rural areas across China was severe, nor were cities immune to it, and citizens paid a heavy price due to it. Even in Beijing, the Capital of China, numbers of people in the urban areas were also exposed to the hidden troubles from underground water. The report provided an in-depth analysis of the source of the problem: the national survey was imperfect, form was greater than substance in various provinces and municipalities, the monitoring capability was extremely underdeveloped, adding difficulties to the controlling of underground water. Based on investigations, the report also adopts the practice from the international Superfund Act and offers a constructive conceptual design.

Best In-depth Report Award: *Ten-year Reflection on New Energy*, Xie Dan, Chen He
The speech from the judging panel of *Southern Weekly* at the award ceremony: In 2012, China's new energy industry underwent a sharp decline from its peak. Xie Dan et al were the first to make a summary of and reflected on the sweeping bubble that had burst in China's new energy industry at?? in early 2013. They interviewed more than 30 entrepreneurs, local officials, decision-makers and scholars to analyze how the bubble grew in new energy, how the opportunities for rescuing the situation were missed and finally how a large number of enterprises suffered losses and went bankrupt. The

original title of the article was "The Fiasco in New Energy," which was the first-ever systematic consideration made by them as media professionals. This view caused great repercussions within the industry.

Best Young Journalist Award: Wang Tao
The speech from the judging panel at the award ceremony: Since 2012, Wang Tao has been the candidate for the Best Young Environmental Journalist Award of the Year. This young man, a new member of the industry, is one of the outstanding people among the new generation practitioners of environmental news coverage in China. In 2013, Wang Tao continuously reported on air pollution and took strong swift actions with sharp observations on issues ranging from severe haze to discussions about the effect of air pollution on human health and also on the issues discussed in An Action Plan for Air Pollution Prevention and Control as well as realistic problems in haze control at the local level. Her report *Haze Swept through the Regions South of the Yangtze River* provided an analysis of the causes for individual cases involving local haze; *Local Haze Control, Lost on Journey* described the awkwardness and plight of haze control in some areas; *The Smell of "Air" in the Political Season of "Two Sessions"* depicted the public concerns and political involvement in the issue of air pollution from the perspective of "Two Sessions", so it was unique among the reports during "Two Sessions"; *Air Pollution Caused Diseases, the Burden was Heaviest in China* was made by opportunely interviewing international authorities, and it took the lead in the reports about air in terms of public health. As a young journalist, Wang Tao kept an active thinking, high output and stable level in 2013.

Best Citizen Journalist Award: Wang Chunsheng (@caixiangzhijian)
A citizen of Weifang City, Shandong Province, Wang Chunsheng has continually investigated and reported on local environmental pollution and has supervised local environmental governance through his microblog. Meanwhile, he has also legally applied to force such enterprises as Lianmeng Chemical, Juneng Special Steel and Chenming Paper to disclose information regarding their pollution activities; he has also applied to obtain environmental information from such enterprises. He has a good, interactive relationship with government departments, including the environmental department in Shandong.

The speech from the judging panel at the award ceremony: When he appeared and revealed environmental pollution in his microblog, he was just one of numerous complaining netizens; when he frequently changed online username and continued to expose pollution, he acted like an indomitable online guerrilla. If he quickly disappeared online, nobody was surprised. However, he has constantly been investigating and reporting local environmental pollution for a long time, and he has legally applied to a number of large enterprises for disclosure of their environmental information, he has a good, interactive relationship with relevant departments of the provincial government. Finally, we fully affirm that he is an amazing citizen journalist; he still delivers wonderful results amidst the decreasing influence of the microblog.

APPENDIX XI

2014 Winners of UNEP Champions of the Earth Award

The United Nations Environment Programme (UNEP) announced the 2014 Award for the UN Top Environment Award which recognized innovators and policymakers who save lives, improve livelihoods, and better manage and protect the environment.

Launched in 2005, Champions of the Earth recognizes outstanding visionaries and leaders in the fields of policy, science, entrepreneurship, and civil society action.

Champions of the Earth award five prizes: Policy Leadership, Entrepreneurial Vision, Lifetime Leadership, Science and Innovation, and Inspiration and Action.

UNEP hold an awards ceremony at the Smithsonian American Art Museum and National Portrait Gallery in Washington D.C. on November 19, 2014. Winners was be honored by UN Secretary-General Ban-Ki-moon and UNEP Executive Director Achim Steiner. UNEP Goodwill Ambassador Gisele Bundchen also attended the awards ceremony as an honored guest.

The winners of UNEP's 2014 Champions of the Earth Award are:
For Policy Leadership
H.E. Tommy Remengesau, Jr. President of Palau for strengthening the economic and environmental resilience of Palau by spearheading national policies to protect biodiversity.
Sixth President of Indonesia, Susilo Bambang Yudhoyono for becoming the first president from a major developing country to voluntarily pledge to reduce greenhouse gas emissions.

For Entrepreneurial Vision
U.S. Green Building Council for changing the way buildings and communities are designed, built and operated.

For Science and Innovation
Sir Robert Watson, Eminent Environmental Scientist for promoting the science behind ozone depletion, global warming and the impacts of biodiversity loss.

For Inspiration and Action
Boyan Slat, Founder of The Ocean Clean-up Initiative for charting new territory in his quest for a solution to the worsening and global problem of plastic debris in our oceans.

Fatima Jibrell, Founder of Adeso (formerly Horn Relief) for building environmental and social resilience amidst war and devastation.

For Lifetime Leadership
Sylvia Earle, Ocean Explorer and Conservationist for developing global "hope spots" to safeguard the living systems underpinning global processes that maintain biodiversity.
Mario Molina, Nobel Laureate and Renowned Ozone Scientist for spearheading one of the most significant climate-related global agreements ever made.

About Champions of the Earth
Launched in 2005, Champions of the Earth recognizes outstanding visionaries and leaders in the fields of policy, science, entrepreneurship, and civil society action. Whether by helping to improve the management of natural resources, demonstrating new ways to tackle climate change or raising awareness of emerging environmental challenges, Champions of the Earth serves as an inspiration for transformative action across the world.
Past laureates have included Mikhail Gorbachev, Al Gore, Felipe Calderon, Mohamed Nasheed, Marina Silva, Vinod Khosla, and many other such exemplary leaders on the environment and development front.
Xun Zhou, the famous Chinese actress and Yue Zhang, President of the Broad Group, won the award in 2010 and in 2011, respectively.

About UNEP
UNEP, established in 1972, is the voice for the environment within the United Nations system. UNEP acts as a catalyst, advocate, educator and facilitator to promote the wise use and sustainable development of the global environment.
UNEP work encompasses: Assessing global, regional and national environmental conditions and trends; Developing international and national environmental instruments; and Strengthening institutions for the wise management of the environment.

About Awards Sponsors
1. Guangdong Wealth
Guangdong Wealth Environmental Protection is a leading supplier of water purifying products and water treatment integrated solutions in China. The company practices a business model that puts social welfare before economic interests, using the concept "let the sky be bluer and the water clearer". The company invests in environmental scholarships for young university students, organizes clean-up operations, and donates tonnes of purifying tablets to tackle pollution in rivers in Guangdong and Beijing.

2. The Smithsonian Institution
Founded in 1846, the Smithsonian Institution is the world's largest museum and research complex. It consists of 19 museums, a zoological park, nine research facilities, and numerous research programs around the globe in the areas of science, art, history, and culture. Its mission is to preserve America's diverse heritage, to increase knowledge and to share its resources with the world. The Smithsonian's work helps to build bridges of mutual respect and understanding of the diversity of American and world cultures.

3. The UN Foundation
The United Nations Foundation links the UN's work with others around the world, mobilizing the energy and expertise of business and non-governmental organizations to help the UN tackle issues including climate change, global health, peace and security, women's empowerment, poverty eradication, energy access, and u.s.-un relations.

4. The Washington Post
The Washington Post is a leading news publisher that sits at the intersection of business and policy. Through its heritage and pedigree, it has been a pioneer at the forefront of objective, credible, award-winning journalism. An audience of influential leaders depends on the insight and analysis of the organization to make critical decisions that impact their business. The Washington Post paves the path for success through technology and engineering by bridging readers from today into the future of journalism. The 137 year old media company is forging new media consumption experiences for our readers across all platforms.

APPENDIX XII

2014 Winners of the Green China · 2014 Environmental Protection Achievement Award

In order to promote the development and revitalization of a green economy, and continuously draw the attention of the whole society to environmental protection, energy conservation and the future living environment for people, the Green China·2014 Environmental Protection Achievement Award has been successfully granted to commend and encourage the enterprises and individuals making contributions to China's environmental protection activities. With the deterioration of the global ecological environment and the increasing energy crisis, the concepts of environmental protection and low carbon have been deeply rooted among the people; going green and environmental protection have become two of the most watched issues in the world at present and will probably continue to be so in the future. The representatives from the award-winning organizations and enterprises have shared their experience and perspectives about going green and environmental protection.

Outstanding Green Healthy Food Award and Outstanding Environmental Protection Leader Award—Yu Xiang, Chairman of Kunming Junxianghao Tea Co., Ltd.
Asia Outstanding Luxury Real Estate Habitat Award—Yan Jun, Manager, Marketing Department, Wuhan Wanda Plaza Investment Co., Ltd.
Outstanding Green Ecological City Award—Dong Ping, Deputy Director, Publicity Department, Party Committee of Luxi County
Outstanding Environmental Governance Engineering Award—Wu Jichang, Director, Fuxian Lake Management Bureau, Yuxi City
Outstanding Green Ecological City Award—Zhang Lei, Executive Deputy Director, Publicity & Planning Center, Fengdong New Area, Xixian New Area, Shaanxi Province
Outstanding Sustainable Development Enterprise Award—Leng Zhenghua, representative, Wuhan Shuanghu Coating Co., Ltd.
Outstanding Environmental Protection Enterprise Award—Lu Yili, Managing Director, CT Environmental Group Limited
Outstanding Sustainable Development Creative Concept Award—Zhuang Yizhou, Chairman of Yizhou Shiji Co., Ltd.
Outstanding Environmental Protection Listed Company Award—Wu Yongkang, Chairman; Wu Yuqun, Chief Executive Officer & Executive Director, Baguio Green Group

Outstanding Environmental Protection Enterprise Award—Jia Yan, Deputy General Manager, Ningxia Tianyuan Manganese Co., Ltd.

Outstanding Corporate Social Responsibility Award—Zhang Dongchu, Chairman, Ningxia Taijian Real Estate Development Co., Ltd.

Outstanding Green Environmentally-Friendly Building Award—Wang Chenglin, General Manager, Wuhan Center Tower Development Investment Co., Ltd.

Outstanding Green Healthy Food Award—Jin Feng, Chairman, Hubei Conglin Agricultural Ecological Co., Ltd.

Outstanding Green Healthy Food Award—Yang Shiwen, General Manager, Hubei Shihua Distillery Co., Ltd.

Outstanding Green Healthy Food Award—Lin Yong, Executive President, Wissun International Nutrition Group Co., Ltd.

Outstanding Creative Energy Conservation and Environmental Protection Concept Award—Ye Guangqing, Chairman, China Lijie Building Decoration Technology (Hong Kong) Co., Ltd.

Outstanding Green Healthy Food Award—Su Dajun, Chairman, Henan Tongduan Industrial Co., Ltd.

Outstanding Sustainable Development Enterprise Award—Feng Haizhong, Chairman, Shanxi Xinzhong Group

Outstanding Sustainable Development Enterprise Award—Deng Jiexin, representative, Zhejiang Lonsen Group Co., Ltd.

Index

aborigines   177, 181–183, 185
Acid Rain   235, 253
Action against Violation of Environmental Law   231
administrative detention of lawbreakers   73, 76
administrative punishment   85
Africa   132–136, 139, 144–145, 150
Agenda 21   18, 125
agricultural products   63, 65, 69, 179, 181
AIDS   134–135, 139
Air Polluters   236
Air pollution control   9, 26, 30–31, 202, 231, 237, 240
Air Pollution Prevention and Control Law   13, 33, 40, 72–73, 80, 243, 271–273
Air Quality Index (AQI)   241, 262–263
Air Quality Information Transparency Index (AQTI)   200, 203
Ambient Air Quality Standard   251, 253, 263
amendment   3, 14, 33, 35, 72–73, 77, 80, 97, 109, 115, 119, 172, 234, 243, 273
Antarctica   189, 191–192
Antarctic marine areas   187
Antarctic marine conservation   187–188, 199
Antarctic marine ecosystem   189
Antarctic MPA   191–192, 199
Antarctic resources   196, 198
anti-corruption   157, 168, 170–171, 174
APEC Blue   24–25, 27–30, 32–33, 35, 241
APEC Economic Leaders' Meeting   25
APEC meeting   24–30, 32, 59, 171, 240–241
arable land quality   242
area-wide environmental quality for noise   254
assuming liability   86

bats   136–137, 140
bearing capacity of energy environment   54
best environmental report award   284–285
biocapacity   120, 122, 124, 126, 129
biodiversity   2, 9, 141, 163–164, 167, 178–179, 195, 198, 257, 289–290

book jackets   268–269
Brazil   176–186

CAMLR Convention   187, 190, 192
capability for disposal   103
Carbon Capture, Utilization and Storage (CCUS) technology   52–53
carbon dioxide emission   41, 128, 208, 226–227
carbon emission intensity per unit of GDP   39
carbon emission reduction   40, 207, 225, 227
Carbon Emissions Trading   232
carbon intensity   46, 208
carbon market   40–41, 206–209, 218
carbon offset projects   207, 209, 218
carbon pricing   207, 209
carbon tax   41, 45, 51, 207, 209
carbon trading market   40–41
car system   55
catch limits   188, 190, 194, 198
CCAMLR   187–189, 191–192, 195–196
CCICED   125, 129–130
central finance   60–61, 64, 66–68
Champions of the Earth   289–290
Changsha-Zhuzhou-Xiangtan pilot project   60–61
chimpanzees   134–137
China Minmetals   154–155, 157
China's financial industry   158
China's investment   152, 177, 181, 183, 185–186
China's national conditions   58, 91, 161
China's NGOs for environmental protection   5–8, 11, 13–15
Chinese dream   25
Chinese investors   152, 177, 181–182, 184, 186
Chinese rich people   147–148
city groups   38
civil society   1–4, 6, 15–17, 19, 21–22, 157, 176–177, 182, 289–290
classified management   64, 70

clean coal technologies 52–53
clean energy 36, 42, 48, 234
climate change 10, 15–16, 36, 40, 42, 46–47, 52, 83, 120, 125, 163–164, 167–168, 171, 174, 180, 190, 199, 207–208, 232, 239–241, 290–291
climate negotiations 239
climate warming 15
clothing 126–127
coal-based polygeneration 52
coal consumption control scenario 39
coal price 51
coal reduction 37
coal utilization 38, 44, 46, 52
Commission for the Conservation of Antarctic Marine Living Resources 187–188
community practice 11
control risks 155
Convention on the Conservation of Antarctic Marine Living Resources (CAMLR Convention) 187–188, 190, 192, 198
conviction 98, 101
corporate diplomats 178
Corporate Environmental Credit Evaluation 229
corruption 156–157, 159, 163, 165, 168, 170–171, 181
cost-benefit analysis 30, 34
credibility 17, 107, 186
credit evaluation 90, 92–94, 229
crime of environmental pollution 97, 102, 104
criminal law 95, 97, 105, 107
criminal penalty 107
crop production 70
crops 65, 140, 257
cultivated land 60–65, 67, 69–70, 81, 235, 242

decision-making 7, 12–13, 18, 20, 22, 24, 26, 53, 90, 92–93, 116, 160, 162, 269
de-industrialization 186
dereliction of duty 95, 97, 105–106
developing tertiary industry 65
disclosure of environmental information 13, 24, 33, 110–111, 113, 118
disclosure of information 3, 14, 112, 207, 271, 273
disclosure upon request 118
diversification of governance subjects 18
Drinking Water Pollution 232–233
Drinking Water Quality 231
dust-haze 253

ebola 132–133, 135–137, 139–140
eco-cities 128
Ecological Advancement 187, 196, 198–199
ecological civilization 42, 45, 91, 106–107, 120, 122, 126, 275–276
ecological compensation mechanism 50
ecological compensation taxes 51
ecological consequences 142
ecological environmental quality 256
ecological footprint 120–122, 124–126, 129
ecological red lines 40, 45, 239
ecological tax 41
ecology of finance 83
economic growth pattern 18, 24
ecosystem approach 188
ecosystem protection 229–230, 239
ecosystems along Yangtze River 241
education 1, 13, 107, 151, 182, 267, 269–270, 282
efficiency of the coal system 47
electricity generation 225–227, 260
electricity subsidy 225–227
emission data 110–111, 206, 235
emission reduction 40, 196, 200–203, 206–209, 218, 225, 227, 230, 239–240, 244, 266
end energy 47
energy conservation 40, 46–48, 53, 121–122, 128, 196, 206, 208, 239, 266, 292–293
energy consumption 36, 39, 41–43, 47–49, 52, 122, 125, 128, 229, 241, 261
energy efficiency 36, 48, 127, 159
energy intensity 42, 208
Energy Mix Adjustment 229
energy production 260
energy-saving benchmark scenario 39
energy security 16, 50, 208
energy supply security 50–51
energy transformation 43

Environmental Assessors   236
environmental awareness   3, 7–8, 21, 269
environmental behaviors   84, 88–89, 93, 111
environmental capacity indicators   38
Environmental Cases Reported by Public   237
environmental civil social organizations   19–20
Environmental civil society   1–4, 6, 19, 21–22
environmental credit risk   89
environmental criminal justice   96
environmental governance   1, 16–19, 113, 276, 287, 292
environmental impact assessment (EIA)   8, 12–13, 18, 74, 88, 110, 112–113, 115, 151, 265, 272, 284
environmental impact assessment report   74, 272
Environmental Impact Report   74, 236
environmental information disclosure   35, 109–111, 113–117, 119
Environmental Information Transparency   235
environmental knowledge publicity   11
environmental law enforcement   29–30, 78–79, 243
Environmental Liability Insurance   234
environmental pollution crime   95, 105
Environmental Proposals   231
Environmental Protection Achievement Award   292
environmental protection law (EPL)   3–4, 13–15, 21–22, 35, 72–81, 83–88, 109, 113, 115, 119, 125, 156, 234, 238, 243, 271–273, 275
environmental protection laws and regulations   81
environmental protection organizations   6, 9–10, 22, 35, 75–76, 93, 119, 201–202, 276, 278, 280
environmental public interest litigation   14–15, 21, 73–76, 86, 88, 275–276
environmental quality for noise   254
environmental regulations   153–154
environmental regulator   85, 89
environmental remediation industry   67
environmental responsibilities   16

environmental responsibility compensation   86–87
environmental risk awareness   90
environmental risk control   89–90, 92–93
environmental risk management   89–93
environmental risk reporting system   91–92
environmental safety brigades   100
environmental supervision   13, 78, 264, 267
Environmental Tax Law   234
Environment- and Resources-specific Trial Court   236
EPA   113
excess production capacity   48, 50
externality   89

fairness   107, 140, 218
farmland quality   256
FDI   150–153, 161
financial environmental risk responsibility system   91
financial institutions   83–90, 92–94, 155, 157–161, 175
financial regulators   83–84, 89–92, 94
financial risk   89
financing   60–61, 66–68, 70–71, 84, 87, 158, 160–161
food safety   16, 70
foreign investors   181–182, 185–186
forest governance   164, 169, 172
forest resources   164, 171, 258–259
friends of nature   5–7, 9–11, 14–15, 35, 75, 163, 200, 268–271, 273–276, 278–279
fuel oil tax   59

G20   170–171
garbage classification   7, 11
Global Timber Trade   168, 174
global village of Beijing   5–7, 11
Good Governance of the Environment   2, 18, 21
gorillas   135, 137
governance capacity   17–19, 163–164
government-centered theory   17
government failure   16
GOVNGOs   5
green consumption   9, 11, 121–123, 125, 127, 129–131

INDEX

green credit policies   159
Green Earth Volunteers   6, 11
green economy   16, 292
green finance   83, 90
green GDP   49
greenhouse gas emissions   167, 207, 209, 227, 289
greenhouse gas voluntary emission reduction   209
groundwater environmental quality   247
Groundwater Quality   233
gulfs and bays   250

hazardous waste   95, 103–105
haze pollution   38, 80, 253
heavy metal pollution control   238
heavy pollution   28, 33–34, 42, 62, 81, 200–203, 243, 263
High External Environmental Costs   236
high seas   191, 197
historical information   116–117
housing   124, 126–128, 265
human health   2, 16, 63, 101, 139, 226, 236, 263, 277–278, 287
hunting overseas   146–147

illegal environmental behaviors   88–89
illegal logging   163–175
illegal timber trade   163–164, 166, 172, 174
imposition of a daily punishment   73, 76
improved soil from other places   64
independent social third-party forces   83
indigenous land management   184
Indigenous policies   183
industrial solid waste   245
industries with high energy consumption   48–49
industry standards   154, 162
information disclosure   11, 14, 22, 34–35, 72–75, 88, 92, 109–119, 156–157, 174, 200–201, 203, 207, 271, 273, 278
information exchange platform   93
in-process information   114, 116
installed capacity of nuclear power   81
Institute of Public & Environmental Affairs (IPE)   10, 14, 35, 110, 119, 200, 203
institution   144, 291
institutional system   83, 91, 275–276

interest coordination   18
International Union for Conservation of Nature (IUCN)   189
Interpretation of several issues concerning the applicable laws for hearing environmental civil public interest litigation cases   76
invasive alien species   258
investigate and transfer   98

Jack Ma's   142–143
Juan Carlos   143

krill   187, 189–190, 192–194, 197–198

landless peasants   184
Latin America   150, 178–179, 186
law enforcement supervision   13, 24, 29, 33
legal hunting   148
local protectionism   19
long-term measures   24, 28, 30
low-carbon green transformation of energy   36
low-carbon technologies   225–226

malaria   133–135, 139–140
management system   16, 24, 34, 83, 90–92, 154, 188, 196, 232
mandatory requirement   73, 76, 268
Marine and Polar Power   196
marine conservation   187–188, 191, 199
Marine Ecosystems   236, 257
marine fishing industry   197
Marine Protected Area (MPA)   189–192, 194–196, 199
marine protected areas   187, 189, 191, 196
Marine Protection   233
market levers   41
market supply and demand   41
maximum quantity of discharge   87
Measures on Open Environmental Information (for Trial Implementation)   109, 111–113, 115–116, 118
mining areas   62, 65–66, 236
mining industry   151–152
mining investment   150–152, 154, 156–157, 183
modernization   17, 19–20, 52, 180

monkeys   136–137, 168
MPAS   189–192, 195–196
multi-subject participation   18–19

national energy situation   260
national governance system   17, 19
natural reserves   257
New Environmental Law   234, 243
new management systems   90
new urbanization scheme   120, 122–123
NGOs for environmental protection   2, 4–15, 19–23, 272
Nineteen Companies Fined by MEP   235
non-fossil electric power   39
non-fossil energy   38–39, 229
non-governmental environmental protection organization   268
non-governmental organizations   2, 16
normative documents   101, 107, 225
nuclear energy   81–82
Nuclear Safety Law   72–73, 82

objective responsibility management   37, 53
oil consumption   41
open environmental information   109, 114, 116
organic pollution source identification   242
outbound investment   150–154, 160–162
outbound mining enterprises   155
overseas investment   159, 161, 170, 183
overseas mining investment   156

plagues   138
PM2.5 concentration   27, 32, 37, 45, 229
PM2.5 pollution   57
point locations   61–63, 81
policy evaluation   34
political commitment   25
pollutant discharge standard   85, 87, 101, 107
pollutant discharging units   74, 85, 87, 272
pollutant emission   33, 52, 230, 236, 240
polluted water quality   246
Polluters Fined   237
pollution control   9, 26, 28, 30, 69, 80, 202, 206, 231, 237–238, 240–241, 265, 277–278, 284

pollution remediation and treatment   64
pollution sites   63
pollution sources   70, 110–112, 119, 200–203, 233, 237, 274
precautionary approach   188
premature deindustrialization   179
prevention and management   90
primary PM2.5 contribution   44
Prince Harry   144
private car   58–59
protected areas   132, 187, 189, 191–192, 195–196
protection of soil environment   69
public awareness about environmental protection   7–8, 88
public benefit activities   3, 14, 21
public health   21, 30, 33, 36–37, 41–42, 44, 47, 70, 133, 200, 287
public interest litigation   3–4, 14–15, 21, 35, 73–76, 86, 88, 275–276
publicity   7–9, 11, 13–14, 35, 107, 279, 282, 292
public participation   3–4, 6–8, 11–14, 17–19, 21–22, 24, 33–35, 72–76, 88, 109, 112–113, 119, 129, 159, 162, 271–273, 278
public-private partnership mode   71
public services   8, 12, 15–17, 19–20, 22, 151
public transportation   57–59
punishment   33–34, 41, 72–73, 76, 85–86, 91, 98, 108, 206, 273

radical scenario   39
rainfall acidity   253
real-time information disclosure   201
recycling   126, 226–227, 268, 279
reduction   15, 28–29, 37, 40, 42, 46, 48, 53, 111, 196, 200–203, 206, 225, 227, 244, 266, 268, 283
regulatory power   84
renewable energy subsidy   225
replacement and cleaning   37
resource depletion   2, 16
resource saving   268
resource tax   41, 51
risk control   89–90, 92–93, 206
Russia   168, 191–192

safeguarding of public environmental rights and interests   8

INDEX

clean coal technologies 52–53
clean energy 36, 42, 48, 234
climate change 10, 15–16, 36, 40, 42, 46–47, 52, 83, 120, 125, 163–164, 167–168, 171, 174, 180, 190, 199, 207–208, 232, 239–241, 290–291
climate negotiations 239
climate warming 15
clothing 126–127
coal-based polygeneration 52
coal consumption control scenario 39
coal price 51
coal reduction 37
coal utilization 38, 44, 46, 52
Commission for the Conservation of Antarctic Marine Living Resources 187–188
community practice 11
control risks 155
Convention on the Conservation of Antarctic Marine Living Resources (CAMLR Convention) 187–188, 190, 192, 198
conviction 98, 101
corporate diplomats 178
Corporate Environmental Credit Evaluation 229
corruption 156–157, 159, 163, 165, 168, 170–171, 181
cost-benefit analysis 30, 34
credibility 17, 107, 186
credit evaluation 90, 92–94, 229
crime of environmental pollution 97, 102, 104
criminal law 95, 97, 105, 107
criminal penalty 107
crop production 70
crops 65, 140, 257
cultivated land 60–65, 67, 69–70, 81, 235, 242

decision-making 7, 12–13, 18, 20, 22, 24, 26, 53, 90, 92–93, 116, 160, 162, 269
de-industrialization 186
dereliction of duty 95, 97, 105–106
developing tertiary industry 65
disclosure of environmental information 13, 24, 33, 110–111, 113, 118

disclosure of information 3, 14, 112, 207, 271, 273
disclosure upon request 118
diversification of governance subjects 18
Drinking Water Pollution 232–233
Drinking Water Quality 231
dust-haze 253

ebola 132–133, 135–137, 139–140
eco-cities 128
Ecological Advancement 187, 196, 198–199
ecological civilization 42, 45, 91, 106–107, 120, 122, 126, 275–276
ecological compensation mechanism 50
ecological compensation taxes 51
ecological consequences 142
ecological environmental quality 256
ecological footprint 120–122, 124–126, 129
ecological red lines 40, 45, 239
ecological tax 41
ecology of finance 83
economic growth pattern 18, 24
ecosystem approach 188
ecosystem protection 229–230, 239
ecosystems along Yangtze River 241
education 1, 13, 107, 151, 182, 267, 269–270, 282
efficiency of the coal system 47
electricity generation 225–227, 260
electricity subsidy 225–227
emission data 110–111, 206, 235
emission reduction 40, 196, 200–203, 206–209, 218, 225, 227, 230, 239–240, 244, 266
end energy 47
energy conservation 40, 46–48, 53, 121–122, 128, 196, 206, 208, 239, 266, 292–293
energy consumption 36, 39, 41–43, 47–49, 52, 122, 125, 128, 229, 241, 261
energy efficiency 36, 48, 127, 159
energy intensity 42, 208
Energy Mix Adjustment 229
energy production 260
energy-saving benchmark scenario 39
energy security 16, 50, 208
energy supply security 50–51
energy transformation 43

Environmental Assessors   236
environmental awareness   3, 7–8, 21, 269
environmental behaviors   84, 88–89, 93, 111
environmental capacity indicators   38
Environmental Cases Reported by Public   237
environmental civil social organizations   19–20
Environmental civil society   1–4, 6, 19, 21–22
environmental credit risk   89
environmental criminal justice   96
environmental governance   1, 16–19, 113, 276, 287, 292
environmental impact assessment (EIA)   8, 12–13, 18, 74, 88, 110, 112–113, 115, 151, 265, 272, 284
environmental impact assessment report   74, 272
Environmental Impact Report   74, 236
environmental information disclosure   35, 109–111, 113–117, 119
Environmental Information Transparency   235
environmental knowledge publicity   11
environmental law enforcement   29–30, 78–79, 243
Environmental Liability Insurance   234
environmental pollution crime   95, 105
Environmental Proposals   231
Environmental Protection Achievement Award   292
environmental protection law (EPL)   3–4, 13–15, 21–22, 35, 72–81, 83–88, 109, 113, 115, 119, 125, 156, 234, 238, 243, 271–273, 275
environmental protection laws and regulations   81
environmental protection organizations   6, 9–10, 22, 35, 75–76, 93, 119, 201–202, 276, 278, 280
environmental public interest litigation   14–15, 21, 73–76, 86, 88, 275–276
environmental quality for noise   254
environmental regulations   153–154
environmental regulator   85, 89
environmental remediation industry   67
environmental responsibilities   16

environmental responsibility compensation   86–87
environmental risk awareness   90
environmental risk control   89–90, 92–93
environmental risk management   89–93
environmental risk reporting system   91–92
environmental safety brigades   100
environmental supervision   13, 78, 264, 267
Environmental Tax Law   234
Environment- and Resources-specific Trial Court   236
EPA   113
excess production capacity   48, 50
externality   89

fairness   107, 140, 218
farmland quality   256
FDI   150–153, 161
financial environmental risk responsibility system   91
financial institutions   83–90, 92–94, 155, 157–161, 175
financial regulators   83–84, 89–92, 94
financial risk   89
financing   60–61, 66–68, 70–71, 84, 87, 158, 160–161
food safety   16, 70
foreign investors   181–182, 185–186
forest governance   164, 169, 172
forest resources   164, 171, 258–259
friends of nature   5–7, 9–11, 14–15, 35, 75, 163, 200, 268–271, 273–276, 278–279
fuel oil tax   59

G20   170–171
garbage classification   7, 11
Global Timber Trade   168, 174
global village of Beijing   5–7, 11
Good Governance of the Environment   2, 18, 21
gorillas   135, 137
governance capacity   17–19, 163–164
government-centered theory   17
government failure   16
GOVNGOs   5
green consumption   9, 11, 121–123, 125, 127, 129–131

INDEX

safety of human settlements   63
safety of the ecological environment   63
safety standards   251–252
scientific decision-making   24
scope of the subjects   73, 88
Scottish hunting clubs   145
sea areas   249, 257
Seawater Quality   235, 249
secondary PM2.5 contribution   44
seizure and attachment   72–73, 76–77
sentencing   98, 101
short-term measures   28, 30
small and micro enterprises   102
social adjustment mechanism   17
social capital   3, 21–23
social resource allocation   16
soil contamination   242
soil environment   61, 69–71, 81
Soil Pollution Prevention and Control Law   69–70, 72–73, 81, 240–241
soil-related legislative work   69
soil remediation market   60–61, 68, 71
solid waste   239, 245
Southern Ocean   188, 190–191, 193, 198
South-to-North Water Diversion Project   248–249, 264
special fund subsidies   66
species protection   238
statistical accounting system   209
stop production for rectification   87
Strictest No-Burning Zone Scheme   237
substitution effect   39
suppressing secondary industry   65
surveys, evaluation and forecasts   19, 35, 87, 123
sustainable consumption   120–126, 128–131
sustainable development   1, 3–5, 15–16, 21, 47, 50, 66, 84, 94, 120, 125, 128, 155, 160, 162, 164, 169, 191, 193, 198, 207, 280, 290, 292–293
sustainable economic development   10, 43
sustainable transport   128
synergistic effect   40, 44, 46

the application of the law   107
the Friends of Nature   5–7, 9–11, 14–15, 35, 75
the ministry of environmental protection   4, 26–27, 29, 32–33, 35, 60–61, 67–71, 73, 75–78, 81–82, 110–111, 115, 154, 201, 229, 244, 253, 265–267, 274
the third sector in a society   16
the web of life   140
third-party environmental pollution management   240
third-party governance mode   71
Three Gorges Project   248
tolerance   98
toothfish   188–190, 194
top-level design   207, 209
top-level treatment design   66
total quantity control   10, 36, 38–40, 42–46, 48–50, 52–53, 85, 207
trade secret   115–116
transcontinental railroad   178–179
transformation of the economic growth pattern   24
transformation of the energy structure   40
transformation of the energy supply   53
transport   28, 126, 128–129, 172, 236
trophy hunting   142, 146–148
tropical viruses   132

uneven geographical distribution   95, 99
United Nations Environment Programme (UNEP)   46, 159–160, 166–167, 289–290
urban air pollution   200, 202
urbanization   38, 68, 120–124, 128, 131, 203
urban waste reduction   268

viruses   133–134, 137–141
voluntary disclosure   118

waste incineration   6, 78, 225–227, 238
Water Pollution Incident   234
water quality   111, 231, 234, 245–250
water resource   45–46
Water Resources Management   230, 232
Water Resources Protection   231
water saving   46
Wetland Protection   232, 237, 257
wildlife conservation   5, 132, 142–143, 148
win-win outcome   18, 24, 30–31
World Health Organization (WHO)   38, 45, 132–133, 232
WWF   35, 120–121, 124, 127, 129, 144, 168

Zijin Mining   152, 154, 156

Printed in the United States
By Bookmasters